W9-CES-777

## *Praise for*

# THIS IS WHAT AMERICA LOOKS LIKE

"Ilhan Omar has a quintessentially American story, told beautifully in the pages of this book. In the House, Ilhan is an eloquent spokesperson for her point of view. I am grateful that her journey brought her to us."

—Speaker Nancy Pelosi

"Congresswoman Ilhan Omar is a fearless, pioneering badass with a titanium backbone, and her memoir is everything she is, in addition to being a rich, witty, crackling good read."

—Dave Eggers, author of *The Monk of Mokha* and *The Captain and the Glory*

"Ilhan has been an inspiring figure since well before her time in Congress. This book will give you insight into the person and sister that I see—passionate, caring, witty, and above all committed to positive change. It's an honor to serve alongside her in the fight for a more just world."

—Congresswoman Alexandria Ocasio-Cortez

"Representative Ilhan Omar is not just pushing America to live up to its best ideals—she's showing us how the struggle for inclusion and solidarity can transform our communities in the here and now. This book is a gripping, wonderfully frank account of a remarkable political journey that is just getting started. As dazzling as its author."

—Naomi Klein, author of *On Fire: The (Burning) Case for a Green New Deal*

"This beautiful book is a glimpse into the life of a woman I admire, respect, and love so fiercely. Ilhan, you reaffirm that we belong everywhere and so do our stories. Thank you for sharing yourself with the world and for the steadfast resolve you bring to the work of justice and the upliftment of community." —Congresswoman Ayanna Pressley

"This book is a treasure! With every beautifully written page, we are given precious insights into the extraordinary experiences, struggles, and mercies that shaped the freedom fighter we know today. Omar has gifted us with a perfect memoir: reflective without being self-indulgent, intellectual but still passionate, pragmatic yet brimming with hope. After reading her story, even Omar's biggest critics will be unable to deny her principled political commitments, visionary leadership, and profound love for the entire human family. During these trying political times, this book is a desperately needed reminder that the type of world, and leadership, that we deserve is closer than we think."

—Marc Lamont Hill, *New York Times* bestselling author of *Nobody: Casualties of America's War on the Vulnerable, From Ferguson to Flint and Beyond*

"*This Is What America Looks Like* is centered on the lived experiences of my sister in service, Ilhan Omar. It is a powerful and inspirational story that demonstrates the need for representation in government that reflects the diversity of our nation."

—Congresswoman Rashida Tlaib

"A truly brave book from an extraordinary American. Any refugee or immigrant who considers the United States their adoptive home is fortunate to have Representative Ilhan Omar as one of our country's essential leaders."

—Jose Antonio Vargas, winner of the Pulitzer Prize and author of *Dear America: Notes of an Undocumented Citizen*

# THIS IS WHAT AMERICA LOOKS LIKE

# THIS IS WHAT AMERICA LOOKS LIKE

## MY JOURNEY FROM REFUGEE TO CONGRESSWOMAN

ILHAN OMAR

*with*

REBECCA PALEY

*An Imprint of* WILLIAM MORROW

 For information, address HarperCollins Publishers, 195 Broadway, New York, NY 10007.

HarperCollins books may be purchased for educational, business, or sales promotional use. For information, please email the Special Markets Department at SPsales@harpercollins.com.

FIRST EDITION

*Designed by Renata De Oliveira*

Library of Congress Cataloging-in-Publication Data
has been applied for.

ISBN 978-0-06-295421-3

20 21 22 23 24 LSC 10 9 8 7 6 5 4 3 2 1

*To*
*Baba*
*Habon Abdulle*
*Isra, Adnan, and Ilwad*
*America*

# CONTENTS

# PROLOGUE

"Thank you for helping to uplift so many girls from all over! Love from Seattle."

"Salaam sister—from the West Side in Senegal to Detroit #13 strong. You continue to amplify us."

I don't remember when they first appeared on the wall outside my office in the Capitol Building: the Post-it notes with words of admiration and encouragement left by people from as far as Duluth and Delhi.

"Congresswoman, traveled from Oregon and HAD to see you! Thank you for being BRAVE, BOLD, AND OUTSPOKEN. If UR ever in Eugene, I can show you around."

"Thank you so much for all you're doing to protect our courts!"

"Keep fighting for immigrants."

I do know when they started to become a problem for Facilities. It was a few months after I became the first Somali American Muslim woman elected to Congress in 2018—right after President Donald Trump began his Twitter attacks against me.

"Ilhan is an American hero!"

"No matter what they say, we'll always have your back!"

Overnight, a large mosaic of multicolored squares grew up around the American flag and plaque bearing my name and the name of the state I represent, Minnesota. Maintenance asked us to remove them, so my staff took the Post-its down and put them back up on a wall inside my office. But visitors to the congressional office building, open to the public Monday to Friday, 7:00 A.M. to 7:00 P.M., continued to put them up.

"Thank you so much for being a voice for those who cannot speak for themselves."

"You are a soldier of the people for peace & justice for all."

They just kept coming, so Facilities gave up. Now the patch of little neon notes is a permanent bright spot on the otherwise austere white walls of a municipal hallway.

What I am most proud of is not the visible expression of support for my work as a legislator, fighting for all my constituents' ability to participate in our democracy. Nor is it the high praise or the depth of emotion—although these sentiments have gotten me through some tough times. What I am most moved by is the incredible diversity of the message writers. They come from different places and perspectives. There are Post-its from teenage girls, who dot their exclamation marks with little hearts, and from Senate staffers, who took a moment from their long days in a cynical city to jot down a positive word.

A blue heart-shaped sticky note with "Republican women support you" was stuck beside a regular yellow Post-it that read, "From one black immigrant to another, please know that I love you." I learned the Hebrew letters on a light blue note, מיר וועלן זיי איבערלעבן, were actually Yiddish for "We will outlive them," which Hasidic Jews in 1939 had turned into a song of resistance in the face of a Nazi commander. That all these people and more would choose to stand behind me—a Muslim immigrant who had arrived in this country from Africa speaking only two words of English—is proof enough that there are stronger bonds than identity.

As a refugee who fled civil war as a child, I am still trying to figure out where I fit in—which is perhaps why the most important note I found sticking to the wall outside my office had only three words.

"You belong here."

# 1

# FIGHTER

*1982–1988*
*Mogadishu, Somalia*

The teacher quickly put a student in charge of my third-grade class before she stepped out of the room. This was not unusual in my elementary school, where students stayed in the same classroom while our teachers for different subjects rotated in and out. When transitioning between periods, teachers usually designated one child to keep the rest from getting too rowdy.

Like all kids, we were prone to abusing this position. Today, though, the boy in charge really let his newfound power go to his head. Almost immediately, he ordered another, smaller boy up to the chalkboard to write an assignment.

"I beg you," said the boy being ordered to the board, "leave me alone."

But the tall boy in charge was determined to humiliate his classmate, who was a minority in every sense. Poor, small, and an orphan, he didn't have the crisp white shirts, ironed uniform trousers, and shiny school shoes of the middle class that the large boy and I both came from.

The big boy continued to taunt his victim, escalating his threats when his classmate wouldn't rise from his seat, until finally he shouted, "*Hooyadawus!*" which means "Go fuck your mother" in Somali.

I burned in my seat. I always hate it when people use vulgar language, but I get really angry when it involves mothers, who I knew from the beginning were sacred—even if I didn't have one. I mean, everybody was always talking about how important mothers are. In Islam, my native country's main religion, we learned that "Paradise is under the feet of mothers." You were supposed to bow to your mother, abide by her every wish, not debase her.

There were also deeper forces at play than my seven-year-old brain could recognize in the moment. Although thanks to my older sisters and many loving aunties I didn't lack for mothering, my mother, my *hooyo*, had died when I was a preschooler. I don't have a single memory of her, even though I remember other things from that age—like family members fighting over whether or not I should start school. Some of my aunties and uncles thought I was too young, because technically you were supposed to wait at least until you lost your first two teeth. "She'll lose her books," someone said. "She won't know where to go," another argued, "and the other kids will steal from her." But I didn't stop complaining until they let me go. And, no, I didn't lose my books or get robbed, even with all my baby teeth intact.

I remember all of that clearly, but my *hooyo*? What she looked like, something she said, even what she died of? Nothing. As an adult, I went to a hypnotist to see if he could help evoke something, anything—a voice, a touch—but nothing emerged. I still find it so odd.

Whether it was an early commitment to my religion's teachings or the fact that an absence can loom larger than any reality, mothers were a big deal to me—and I didn't like anybody to disrespect them.

"He's not going to get up," I said to the bully. "You're supposed to make sure nobody gets out of their seat, not give us assignments. So you're just going to sit and shut up, and we're going to wait for Teacher."

The boy, at least two heads taller than me, was not impressed. "If *you* don't shut up, you'll be sorry," he said menacingly.

I was a particularly tiny child, so anyone who didn't know me assumed I was a coward. The runt who always got bullied at school. But I wasn't afraid of fighting. I felt like I was bigger and stronger than everyone else—even if I knew that wasn't really the case.

"I'll meet you in the rear courtyard after school," I said. That was the place where all the kids went to fight.

Right before the next teacher entered the room, the boy who I had stood up for whispered to me, "After school, I'm going to run, because after they beat you, they're going to beat me."

"If you don't want them messing with you every day," I replied, "you've got to stand up for yourself."

He might have been a wimp, but he was no liar. When school let out, he kept his word and ran.

With a crowd of kids screaming around us, the bully and I began fighting. I was small but a good fighter. I pulled the boy down and rubbed his face in the sand. When my brother, Malaaq, who was in the eighth grade, arrived to watch the fight and saw me grinding the boy into the ground, he shouted, "Ilhan! What the hell?"

My brother wasn't actually surprised to see me at the center of a fight, just annoyed. There was always a slew of parents coming to our house to complain that I had hurt their children. My dad would just laugh. "The only child nobody should be coming here to complain about is my smallest baby."

**YES, I WAS THE BABY OF A LARGE FAMILY, AND YES, I WAS SMALL.** But that had nothing to do with the sticks growing in the bushes by the gate outside our house, which were perfect for beating back any kid who chased me home from school. I had the independent mindset of an only child. I didn't feel young, in no small part because I was never treated like a child. No one was patronized in my brilliant, loud family.

In our Mogadishu compound—filled with African art, books of history and Somali poetry, and music—the disagreements were constant. We were a multigenerational family—aunties, uncles, cousins, and siblings from my maternal side, all living together.

We were unlike a traditional hierarchical Somali family, where when the father or mother spoke no one else dared utter a word. Instead, everyone, even the youngest child, me, was brought into every decision. Sometimes I wished Baba, my grandfather, and my *aabe*, my father, would take on more authoritarian roles. They were annoyingly accommodating to each person's opinion and patient during the ensuing arguments. Everybody was always screaming about what we should do, even when it came to what we were going to eat for dinner. The constant conflict made us at once close and distant from one another. Despite our differing points of view, we all were accustomed to disputes—we had that in common.

There was nothing typical about my family. To this day, I don't know a family quite like ours. But in Somalia, where members of an extended family living together are almost always patrilineal, we especially stood out, since my *aabe* had moved in with my mother and her family after they were married. Sons usually assumed responsibility for supporting their parents as they aged. Hooyo, however, wouldn't agree to marry Aabe unless she could stay with her family.

My father didn't have a full appreciation of what he was getting into when he decided to leave everything he knew by the wayside for love. Although I would perceive different conflicts in him when I grew older, as a child, the greatest one I noticed had to do with his diet. Aabe, who won't touch seafood, married into a family where fish was the primary source of protein. Furthermore, although he ostensibly fit into my mother's world, as is often the case, he could never completely forget the ways he was born into. My father, an educator, came from a traditional patriarchal Somali family where the boys, the primary beneficiaries of educational investment, were raised to become the leaders of their future families.

Meanwhile, when my grandfather welcomed his firstborn child, my mother, he promised himself that she would be treated the same as, if not better than, any male firstborn. Custom dictated that only the birth of a boy was a moment for pride. But Baba, who had a huge presence, was nevertheless *very* proud. He was opinionated and sure of himself, but not without reason. He had one of the sharpest memories of anyone I ever met. Well-read, he had the knowledge of so many books at his fingertips. When he wasn't working at his government job, helping to run the country's network of lighthouses, he liked to fish and play cards. Baba was also a great cook. He was a purist when it came to the ingredients he used to prepare dishes of his specialty, Italian cuisine. His minestrone was my favorite food.

Just as he wouldn't compromise on the quality of the tomatoes in his soup, Baba didn't waver in his convictions. He stayed true to his vow to raise his daughter as an equal to his sons. When Hooyo met Aabe, she was in her twenties, which was very rare at the time, since women predominantly married in their late teens. Not only that, but she was also gainfully employed as a secretary for a government minister. I don't know that my grandfather needed her financial support, but my mother had a sense of duty about living up to the responsibility and unusual privileges she had been afforded by her father.

Everyone knew that if you ever needed Baba to sign on to something or calm him down about a dispute, you needed to talk to his daughter. She was my grandfather's true confidante. I wasn't surprised by the stories I heard time and again about how while she was alive, whatever Hooyo said, went. That's because Baba continued to invest a lot of time and energy in the girls of the family (more than he did with the boys, according to my uncles). He was extremely close to us and did not adopt the traditional patriarchal role of the protector that Somali men usually fall into with the opposite sex. He treated us as equals.

It's always hard to say why a person goes against cultural norms. My grandfather's freethinking partly stemmed, perhaps, from the fact that he didn't come from one of the country's formalized clans. The maternal side of my family was Benadiri, a Somali ethnic minority who trace their lineage to Persians, Indians, and Bantu peoples from West Africa and Arab Yemenis. Successful traders credited with helping spread Islam to Somalia, they settled in port cities like the country's capital Mogadishu, where my grandfather was born and raised. I think Baba embraced the idea that if you don't fit in anyway, you might as well do what you want.

**THE ONLY PLACE WHERE I COMPLETELY FIT IN AS A CHILD WAS** within the walls of my family's compound. Otherwise, I wasn't quite enough of any one thing. Although officially I belonged to my *aabe*'s clan—one of the most powerful in the country—I wasn't fully Somali because of who my mother was. Not that anyone, other than our neighbors, really was aware of this, since we weren't stereotypical Benadiri, known for their light skin and passive natures. Many of my aunties and uncles, as well as my grandfather, had darker skin like me. And no one in our family was remotely passive.

As the youngest, I was spoiled, but then again I really wasn't. Our family of civil servants and teachers was well off enough to have a guarded compound and driver. But I didn't like the attention I received from the other kids for the in-your-face privilege of our white Toyota Corolla and our driver, Farah—nor the constraints. I hated being driven back home after school and usually tried to walk, which meant trouble for Aabe, since that's when the fights with other children took place.

I also wasn't enough of a girl, at least in the traditional sense. None of the women in my family were expected to cook and clean—like most Somali women. We certainly had just as many, if not more, opinions than the men in the house. But I also did what boys did outside the house. I played soccer. I climbed trees. I snuck into the movie theater. No other girls I knew did any of that.

My tomboy ways only fueled the talk among the neighborhood women about "poor Ilhan," a girl growing up without a mother. Never mind that I had all the love and attention of a crowd of caring adults, they reasoned, I must have been deprived of a mother's affection and guidance.

There were so many assumptions about who and what I was supposed to be, and none of them fit the description I had of myself. But I wasn't burdened by the discrepancy. Indeed, I never bothered to answer for it. Instead, I followed Baba's example. If there wasn't a world out there to fully embrace me for who I was, I didn't have to worry about appeasing anyone.

# 2

# WAR

*1989–1991*
*Mogadishu, Somalia*

I was eight years old when civil war broke out in Somalia. One day everything was okay, and the next, there were bullets piercing not only buildings but also people.

Bullets raining from the sky were a constant. Sometimes, though, the fighting got too loud even for us who had grown accustomed to it. During those times when the adults worried that the militia might be closing in on our neighborhood, we fled to our great-grandmother's neighborhood. As the gunfire and rumors intensified, at least we weren't sitting ducks. In the reality of war, sometimes running for shelter somewhere else makes you feel safer—even if it isn't so. On the way to my great-grandmother's home, I saw bodies piled up on the street. We stepped over them.

The adults didn't know what was happening, even though I felt they should. Instead, I kept hearing them say the same thing over and over: "I don't understand how everything just turned."

I remember everything shutting down. School was the first institution to go, but eventually the mosques, the postal service, the television stations, even the market closed down.

The city's major outdoor market was just to the right of our home. On the other side was a main thoroughfare. On a normal day before the war, it was a bustling spot with people and cars coming and going all the time. During the war, the activity was very different.

The Makka Al Mukarama, a nearby hotel owned by Ali Mahdi Muhammad, the man who would name himself Somalia's new president, became his headquarters. As mortars and bullets flew from one side of the conflict to the other, they went directly over our house. The noise was almost constant, and they lit up the sky overhead at night. They kicked up dust all day when they hit the ground or concrete buildings. Our house was hit multiple times, although, thankfully, no one was ever hurt.

On the rare occasion when the ammunition and its racket would stop, we became uneasy. What was next? Once, during an unusual moment of quiet, someone opened the door to our compound and I spotted a mother walking down the street with her child. "What is she doing?" I wondered. "Where could she be going?"

What was once unremarkable—a mother and child walking down the street, a sky free of bullets—was no longer normal. That sight became a symbol of everything. Nothing made sense anymore.

The chaos in the streets meant access to food was limited. We didn't have vegetables or meat, but as my family members reminded me when I complained about the constant meals of beans, we were the lucky ones. At least we had something to eat. Somalis were literally starving. The market was closed, and food supplies were being hijacked by local militia. In 1992, the year after the war began, an estimated 350,000 Somalis died, many from disease and starvation.

My family got very creative in figuring out new dishes to make out of the same staples of stored beans, rice, flour, sugar, and oil. We ate rice ground into porridge for breakfast and plain rice for lunch, and skipped dinner. (To this day, I hate plain rice. It brings back that time when everyone smelled like a bag of rice. It seeped into people's pores like we had drowned in it.) On occasion, if there wasn't too much fighting around us, we would eat bread baked with halwa. As I bit into the gelatin sweet of sugar, cornstarch, oil, and spices tucked inside the homemade bread, I felt incredibly fortunate that my aunties and uncles could make such tasty things. Mostly, though, there were kidney beans and rice. A picky eater, I was never asked to consume them without at least a little bit of oil and sugar on top.

That this strange life was our new reality didn't fully sink in. One day I had everything I needed, so much joy. The next, it all vanished. I was supposed to get used to that? As a child, I found the adjustment hard.

I ALSO FOUND THE WAR ITSELF CONFUSING. IN OUR HOUSEhold, where everything was discussed in the open, the adults weren't hiding in rooms to talk about the situation. All the time, different family members debated what was actually happening to our country and how it would all end. As is true of any political crisis throughout history, where everyone has an idea of how to fix it and yet nobody can actually fix it, my family argued over every detail of our rapidly devolving world. But it afforded the children an awareness of what was taking place.

As I understood it, a dictator—Mohammed Siad Barre, who had controlled the central government for more than two decades—had been ousted in January 1991 by the United Somali Congress (USC), the Somali Salvation Democratic Front (SSDF), and various other clan-based armed opposition groups. That was exciting for everyone, because nobody wants to live under dictatorship. But by November of that year, the clan factions within the coalition took up arms against one another and soon plunged the country into anarchy.

The battle was between two commanders: Ali Mahdi Muhammad and Mohamed Farah Aidid, who both claimed to be Somalia's rightful leader. The generals were from the USC *and* the same clan, the Hawiye, but they claimed differences stemming from their sub-clans. To most Somalis they looked, sounded, and acted in very much the same way. Each got poor young men from his sub-clan to join his militia with offers of food, machine guns, and *khat*, a plant that acts like an amphetamine when chewed.

Although the Benadiris (my mother's side of the family) were not caught up in this clan-based conflict, the Hawiye systematically targeted my father's northern-based clan, the Majerteen, who had been originally subjected to Barre's brutal crackdown and were the main members of the Somali Salvation Democratic Front opposition group. I remember hearing Aidid on the radio, speaking at a rally where he said his plan for the country was to eat the banana and throw away the skin. I knew exactly what he meant by "the skin," and it included my family.

The majority of our neighbors in Mogadishu were members of the Hawiye clan. I had gone to school with them. They were even our relatives through marriage. I myself would eventually end up marrying a member of this clan that wanted to eradicate mine. It was no different than what happened in Nazi Germany, the genocide in Rwanda, or many other times and places around the world where one day you're all family and the next day some members no longer have the right to exist.

During this clan-based cleansing, we had to hide our identities—which was yet another confusing aspect of the war for me as a girl. I had always learned to take great pride in my lineage, so all of a sudden to have adults furiously insist I never mention it again was bewildering. There wasn't a lot of explanation other than "You will get killed if you tell people."

One day my brother and I went out to take a message to one of my mom's cousins who was leaving the country. In preparation for this errand, we rehearsed what we should say if anyone stopped us and asked about our clan. Baba thought that the two youngest children in our family would be the safest on the streets.

Our cousin didn't live far, just on the other side of a dried-up reservoir, but when we got to the lip of the basin and were about to descend, a handful of militia members shouted at us from a checkpoint, "Stop where you are!" These checkpoints were at almost every intersection, as if to replace stoplights, and were usually manned by men in their late teens to early twenties.

"What are you?" said one of the armed men in uniform who had approached us.

I responded first, because I was afraid the mere fact that my brother was a boy would get him in trouble. I repeated what I had been taught to say, which was my great-grandmother's clan.

"What of *that* are you?" the man asked, referring to my family's specific lineage. That, I had not rehearsed. I wasn't about to ad-lib a sub-clan only to have men with guns realize I was a liar. Sweat broke out across my brother's face.

"That's what I know," I said. "Our family doesn't know their clan. We don't talk about it."

"Well, you can pass now, but don't expect to come back another time without being able to tell us who you are," he said, eyeing me suspiciously. "Make sure you ask your family before you leave your house again."

There was no need, because that was the last time we were allowed to leave our house. My grandfather's notion that being children would protect us from harm had been destroyed along with every other system and institution we had thought permanent. As the war went on, there was no more news, no school, no mosque, no hospitals. Everything was taken over and repurposed into make-shift dwellings for those fleeing violence.

**STILL, IT TOOK A DIRECT ATTACK ON OUR HOME TO MAKE BABA** understand we were no longer safe anywhere in Mogadishu. In all the conversations among my family members, there had been a feeling that—as long as we kept our heads down and didn't cause any trouble—we would be okay. My grandfather in particular was oblivious to the shift taking place within the community he had lived in for decades. He found it hard to imagine that his neighbors—the mothers he had gone to fetch the midwife for when they went into labor and the babies born whom he watched grow into parents themselves—would turn on us.

It was evening; we had just eaten dinner and were sitting around in the courtyard when we heard the sound of clanging metal. My uncle ran to the gate at the front of the house and shouted, "Someone's climbing and trying to get in!"

Almost immediately there was a shuffling noise near the windows. We looked, but there was nothing there except fear spilling out into the darkness.

"Everyone go into the main building of the house," Baba said, "and make sure every door is locked."

We went in, bolted all the doors and windows, and huddled in the large inner room within the main house. There we waited. Some of the adults paced the room. Nobody spoke too loudly, even when they teased my uncle for bringing his gun inside the house where there was nowhere to shoot.

"What are you going to do? Break the ceiling?" His brother laughed.

"If you were going to scare anyone other than us, you should have shot your gun when the door was open," an auntie whispered.

My uncle didn't need anyone to tell him he wasn't prepared to defend our family with a gun he hardly knew how to use. He was obviously terrified that people were going to get in. We all were.

After about an hour of quiet, we heard a massive volley of gunfire at the gate. Bullets—loud, loud bullets—banged over and over at our gate. I imagined metal spraying our beautiful blue steel gate but that it wouldn't budge. The tops of the courtyard walls were lined with broken glass that would cut anyone who dared try to climb over.

Eventually, though, a structure broke loose, and we could hear that they were now inside the outer courtyard. It sounded like an army of men, shouting, shooting, but who knew how many of them there actually were.

It was bizarrely calm inside the house. We all moved really slowly as we found hiding places under beds and inside cupboards. My grandfather made sure nobody's head was in a place where it could get shot.

The men outside set upon the car in the courtyard, stripping the Corolla of anything of value, including its wheels. Then they tried to get in the house. They banged on the doors, broke the glass in the barred windows, and shot at the cinderblock walls. But the house proved impenetrable.

Through a broken window, still protected by bars, we could hear two of them talking to each other, brainstorming ways they might get inside. My youngest aunt and my older sister, who are not far apart in age, happened to be hiding in that room and recognized the voices—they belonged to boys with whom they went to school.

"Jimcale, is that you?" my sister said.

"Yes," he said. "Who is that?"

"It's me, Iskufilan, your classmate!"

"You are going to die today."

His chilling words didn't stop my sister. "Does your mother know you're here, trying to kill us?"

He didn't answer.

"You remember the day I gave you money to buy ice cream?"

My aunt was having a similar conversation with another one of the boys, reminding him of how she would stop by his house from time to time for a visit. My aunt and sister continued to chat amiably with the boys, who were still pretending they didn't know them.

As my aunt and my sister were trying to engage the attackers in conversation, my uncle with the gun tried to fire his weapon out the window. But he hit the concrete ceiling instead, causing cement to rain down all around him.

I was hiding in the bedroom Aabe and Hooyo had shared when she was alive. I was under the bed with my oldest auntie, my *habaryar* Fos, who had become the only mother I ever knew. She was so terrified that she was crying and shaking. I had never seen such a physical reaction to fear in my life and never would again. I tried to calm her down by talking to her. "Close your eyes and pretend that it's just a movie," I said. "Then it won't feel real."

Eventually the noises in the courtyard and the talking

by the windows stopped. We ventured out of the main house. The men had looted our front courtyard, taking everything except the shell of the car, presumably because they couldn't move it.

The men were gone. But Baba knew they would be back. We had to leave. So the following morning, we fled our home—its books of Somali poetry, its collections of African art and music, as well as the safety of its walls. Baba and two of my uncles went to my great-grandmother's house. I didn't get to say goodbye to her, because the rest of our family immediately embarked on the journey to get out of Somalia.

# 3

# WHEN THE MOUTH OF THE SHARK IS SAFER THAN HOME

*1991–1992*
*Mogadishu to Kismayo, Somalia,*
*and Utange Refugee Camp, Kenya*

Early that morning, we boarded a cattle truck bound for a nearby town that was not under the United Somali Congress's control. There was a full load of human beings, maybe thirty or more of our neighbors, squashed into the open bed used to haul animals.

It was the most uncomfortable ride of my life. The truck seemed to inch along, making what was undoubtedly a string of roadside extortion stops. Meanwhile, in the back, we were sandwiched together in the scorching heat. No one could stretch their legs. I was so tiny that I went unnoticed by an older woman who lay down on top of me. At some point, unable to feel my legs anymore, I cried to my *habaryar* Fos, who screamed at the woman that she was sitting on a child. But the woman wasn't paying attention; she didn't care. Everybody was devastated. God knows what story the woman had.

My family had split up and loaded ourselves as human cargo into several cattle trucks that left at different times. Aabe had left first, accompanying my two eldest siblings, who were most vulnerable to the vicissitudes of war. At that moment, able-bodied young men posed the greatest threat and therefore were the first target.

The truck that Fos and I were traveling on seemed to make a million stops. Suddenly, out of nowhere, we heard massive gunfire and men screaming. Our trip abruptly ended. Chaos immediately broke out inside the back of the truck. People crawled over one another without regard, scrambling to get out without getting shot. We jumped out of the truck, forgetting to take the few things we had brought with us from home. I was barefoot as we ran as fast as we could toward the makeshift border with the sounds of weapons firing at our backs. The sun was beating down and we didn't have water, but we kept going.

By the time we made it to our first destination, a

coastal town named Baraawe, night was falling. With nothing to eat and nowhere to go, we were so tired we couldn't think or see straight. We simply fell asleep on the beach. I didn't mind. It was quite peaceful on the soft sand, under a canopy of stars.

With the morning came the worries. Unsure who among our family members had made it to our final destination, Kismayo, Fos began making inquiries among the other people huddled on the beach.

"They are no more," one woman said after hearing a description of my father and siblings. According to her, they had all been murdered.

Oh the stories we heard! My father and brothers—slaughtered. My sister raped. Shocked, we didn't want to believe any of the conflicting reports. But on our second night on the beach, as I went to sleep, I tried to prepare myself for life as an orphan. Will I be mistreated? How will I survive without Aabe?

At dawn on our second day in Baraawe, the ocean air was heavy and sweet. I could hear the rooster's crow. It was that moment in the day when the sun is forcing its way in—a red-yellow orb rising up from the sea. I felt like I heard my dad's voice. I wanted to go follow it. I got up and started walking to where his voice was coming from, and toward the end of the stretch where everyone was sleeping, there he was. Just standing there. I went and put my hand on his face, just to make sure he was real. And he was.

He wasn't dead or mutilated or hung from a tree or anything else I had heard about him. He lifted me up into his arms and we walked back to my aunt.

That moment, it was everything. My father had made it to Kismayo with relative ease. Arriving there, though, he realized that my aunt and I would be defenseless in our journey and so selflessly backtracked to find us and make sure we all survived the journey. Thirty years later, it still makes me tear up to remember this moment.

**WE FOUND ANOTHER TRUCK THAT TOOK US TO KISMAYO, WHICH** was just as much a battlefield as Mogadishu—if not worse. The only difference was that my father's clan was safe there. So there was less fear of being dragged out of our house in the middle of the night to be killed, but more fear of dying from a flying bullet.

In the months we stayed in Kismayo, there were times when the fighting grew so intense that the locals evacuated the town for the airport, where they stayed until the unrest died down. We followed them, but even at the airport, sometimes the violence was sudden and random. Bombs meant for rebel targets sometimes missed the mark and landed on a house, killing everyone inside. You never knew if your house would be next.

Food blockages meant that we might go days without a meal, and when we did eat, we had only the tiniest portions to share. Kismayo was the beginning of my knowing real hunger. I didn't want to eat even when there was food, however, because I was too sad.

Having left a lot of our family behind in Mogadishu, we didn't know who was alive and who was dead. Every week, different cousins or aunties arrived in Kismayo, but my grandfather hadn't made it out yet and I worried about him a lot. The family, which had already sold jewelry and other possessions, didn't have enough money to get everyone out at once. The priority had been to save those of us in the most danger. That was my father and his children, because his clan was a target. So we went first with a few of my aunts, who came along to provide support. Baba believed he could withstand the horror to come. "They are not going to kill an old man," he had reassured me before I left.

But one day he, too, made it to Kismayo. Eventually everyone made it. But the militia persecuting my father's clan took over the city at least once while we were there—we simply had to keep moving, across the border and into Kenya. But for that, there had to be another marshaling of resources and dividing up of our family—this time by physical ability.

One way to cross into Kenya was by foot. Although Aabe and most of my other relatives took a boat, a few of my siblings walked the five hundred kilometers to the border. Baba and I didn't have the strength for the journey by foot or by sea. Neither did Fos, who by this point was pregnant. So we were smuggled out of the country at great cost on a small plane used to bring in contraband shrimp. My aunt and I traveled in one aircraft, and my grandfather in another. The smell of putrid seafood was horrific, and the trip was extremely shaky. As I watched other passengers throw up, I wondered if it was motion sickness or the smell.

**WHEN WE ARRIVED IN KENYAN TERRITORY, WE DIDN'T LAND AT** an airport but in the middle of the desert, where we got off the plane and started walking. Whether it was exhaustion or relief at arriving in Kenya, all I remember of that day is sitting in a field surrounded by other Somalis as tired as we were and being handed numbers on pieces of paper by staffers of relief agencies.

At first it was just Fos and me. The Utange refugee camp, five kilometers west of the port of Mombasa, was hot and bare. In the camp, there were no structures, just red dust with the sun beating down from a too-bright blue sky overhead. Looking around at the empty land, I was confused. This was where we were going to live? But how? I didn't ask any questions, though. I could tell my pregnant aunt was tired and unwell from the journey.

She also had business to take care of. We had to pick out a tent and make sure that we reserved space for the rest of our family, who were still on their way. As my *habaryar* gave the names of my father, siblings, aunties, uncles, and cousins, I was reassured that their arrival was official.

Despite our bleak surroundings, I didn't feel like something bad was happening to us. There seemed to be a lot of concern as to how we were doing; we were given some food and things to drink. The real reason I felt everything was going to be okay, though, was because I had Fos by my side.

Whenever people back in Mogadishu pitied me because my mother was dead, I didn't understand it. Not only had I been so little when she died—I didn't remember her or the pain of losing her—but I also had my aunt, who made sure I had everything I needed.

Habaryar, which translated from Somali means "small mom," was different from my other family members. Soft-spoken and calm, she was never in conflict with anyone, never argued, and rarely said no. Her personality and presence were easy. She got things done without making a fuss—and she did everything with coiffed hair and in heels. I can still hear the sound of her heels clicking on the floor of my elementary school, where she had been a teacher. Although she taught math, she used music and dance, both of which she loved, as part of her lessons. She was a super human, but one who didn't need her powers to be recognized or celebrated.

She was the person I knew would make sure that if I didn't like the food served at a meal there was going to be something else I would be happy to eat. Fos was the person watching out for me, even if nobody asked her to and she never talked about it.

And in her eyes, I could do no wrong, which for a child who was constantly getting in trouble was a minor miracle. Even my father was constantly reprimanding me. I got in trouble with him. I got in trouble with Baba. I could get in trouble with just about anybody. But I couldn't get in trouble with her. "This is my sister's baby," she would say. "Everybody leave her alone." To know there was someone whom I could always count on but who also let me be whatever I wanted to be gave me more security than anything else could.

**WE HAD ESCAPED FLYING BULLETS OVERHEAD AND STARVATION-**level food shortages, but I soon discovered that the Utange camp had unseen dangers of its own. Malaria, dysentery, and respiratory diseases were rampant. The camp lacked sanitation and healthcare, and the refugees living there were malnourished and under great psychological stress.

A couple of weeks after we arrived, Fos contracted malaria and grew extremely ill. There was no medicine except for ancient remedies practiced by the other refugees living in the camp. I went all over to find help for my aunt—I tried everything I could think of to make sure she did not die.

By early 1992, my father, grandfather, and most of

my other relatives were in Utange. Nevertheless, my aunt grew sicker and sicker until she could no longer get out of bed. To my ten-year-old mind, my *habaryar*'s illness was longer and sadder than anything I had ever known.

Early one morning, though, she woke up and had so much energy it was as if she were her old self. Fos talked to me, telling me how I needed to calm down and not fight with everyone all the time. She begged me to have patience with my siblings and my father. I was so happy to see her awake and alert I didn't even care if she was uncharacteristically scolding me.

"It's nice that you're able to fully talk," I said.

"I'm fine," she said. "Everything's going to be fine."

Now that she was feeling better, my aunt told me she was strong enough for a burning remedy to get rid of the toxins. So I set off through the camp in the early morning hours to find the lady who practiced it. When I found her, she was making breakfast for her children and told me to come back later. I begged her to come right away, but she said she couldn't and promised she'd be by later.

When I returned to our tent, Fos was lying down again and fading in and out of consciousness. As unexpectedly as she had regained her strength, so had she lost it, and now she was worse than I had ever seen her. She wanted somebody from the mosque. I ran to the next tent over and woke up my dad, who came right away. My aunt asked him to recite Al-Baqarah, a chapter from the Quran, and told me to get her father. But my dad said that Baba was in the men's bathroom, so I wouldn't be able to get him.

"No, you continue reading the prayer. I don't know if I'll make it," my aunt said, "and she doesn't have a problem going into the bathroom."

She was right. I ran straight into the communal bathroom, a long row of men squatting over an open trough of sewage who were shocked to see a girl right in front of them. I screamed my grandfather's name until he came running.

Back in the tent, Baba read the Quran to my aunt. For maybe five minutes, certainly no more than ten, I watched as my grandfather's face grew sadder and sadder. "The spirit is leaving her feet," Aabe said in an effort to comfort him. "Don't be sad. It's a transition we all make."

Neither of them realized I was still standing in the doorway of the tent listening the whole time, until they went to cover my aunt's face. When he saw me, Aabe yelled at me to get out, and I did—not because he was angry with me for watching her die but to run and tell the others that she was gone. Crying, I woke up the rest of our family, which confused and alarmed them, because I never cried.

I don't think I've known greater devastation and sorrow than when Fos died. The experience was profoundly difficult not only because she was the first person I watched die and because she was pregnant. It wasn't just that she was a mother figure to me. My aunt's death meant in very real terms that there was no such thing as escape in this life. If you are destined to die, it doesn't matter how far you travel—you will die. Nothing is permanent, and that fact made me really angry.

When I later relayed my feelings to my father, he said, unfortunately, they were all true.

"All you can hope for is that you die surrounded by the comfort of the people you love," he said. "If you're a good person, you won't die alone."

# 4

# REFUGEE

*1992–1995*
*Utange Refugee Camp, Kenya*

Every week, somebody else died. I felt like I witnessed more death in the refugee camp than I did during the war. I had a hard time making sense of it. All these Somalis invested so much of their energy and so many resources in getting themselves and their children to safety, only to find themselves newly in danger. Flying bullets were no longer a threat, but there was illness and starvation.

You could see the calculus visibly malnourished parents made to feed their children first. Kids were constantly being orphaned. I would play soccer with a boy one morning, and the next he would be an orphan. It was an everyday occurrence. Extended family or clan members took care of these children, but it wasn't the same as having your parents.

A family with six children, distant relatives of ours, lost both their mother and their father in the span of two weeks. The youngest, Umi, was no more than six months old, and the oldest, Naj, was maybe eleven. It immediately became a communal responsibility to figure out how in the world these kids were going to make it. Everyone took turns helping out. Even I would get sent to help with baby Umi. Holding her, I watched her big sister Naj struggle to keep it together.

None of us held out much hope for the six siblings, which is why I was stunned when one of them reached out to me, decades later. After I won my congressional race in 2018, the middle sister, Amina, called my district office. She told me that Naj had stayed in Africa, but that the rest of the siblings had resettled in Minnesota, Iowa, and Washington State. Amina, who lived in Minneapolis, wanted to know if people needed tickets in order to get into my upcoming inauguration. "I want to come see you," she said. "My little sister wants to come, too."

"Wait. Your little sister? The one who was a baby in the camp?" I said. "She's alive?"

"Yes! Umi is doing great. She and her husband, Ryan, live in the Midwest. She heard about your swearing-in ceremony and wants to come."

Although I hadn't actively thought of Umi in years,

I realized that I assumed she had died in Utange. There must have been some part of me that still didn't believe she survived, because when Amina introduced her at the ceremony I held in my office, it didn't register.

"I'm the baby," Umi said.

I stopped and looked at her. She was so beautiful I started to cry. There were a few individuals from our time in the camp that I was certain wouldn't make it out, and she was one of them. In fact she had more than made it out—clearly this vibrant, smiling woman was thriving to boot. It restored my hope that while nothing is permanent, the universe sometimes does take care of people.

**WHILE I LIVED IN THE CAMP, HOPE WAS IN SHORT SUPPLY. THE** best you could ask for was survival. I didn't worry about myself as much as I worried about my dad, whose immune system wasn't strong enough to protect him from malaria. He got sick all the time, sometimes so violently that he passed out. The truth was everybody I knew got sick—that is, everyone except for Baba and me.

I didn't catch anything except chicken pox. Without any kind of remedy or medicine, my skin burned under the scorching heat. I could literally hear the blisters popping. Despite the physical agony I was experiencing, I knew chicken pox wasn't going to kill me. All of my siblings, my cousins, and my aunts and uncles had malaria.

And when someone fell ill, we never knew if they were going to recover.

For the first year and a half in the camp, my grandfather and dad walked around like zombies. All the adults were like shells of humans. If I worried constantly about Aabe dying and becoming an orphan, I can't imagine how he must have felt. As a nine-year-old, I didn't think about my father's anxieties. Instead, his shell-shocked attitude registered as a lack of presence on his part. If I pointed out that someone we knew had been taken to the Red Cross clinic by wheelbarrow, as all sick people were carried, he would get a vacant look on his face. It was an expression that I had never seen before, as if nobody was there. And I didn't like it. At first I was frightened, but eventually I grew inured to it.

There were a lot of moments like that, and not just with Aabe. I liked to sit next to Baba as he read the Quran and would recite it with him. But sometimes he just stopped and gazed out at nothing. There was nothing to look at except a dusty expanse, baking in the sun. It didn't matter, because in that moment he was gone like my father, both men consumed by their inability to do anything to ensure that we would survive.

They were overwhelmed by helplessness as they considered that the children they had imagined bright futures for now had nothing. Aabe, Baba, all the parents in the camp shared this common despair, although they barely spoke to one another. Each was drowning in a sea of the unknown, alone.

I don't talk to my dad about that early time in the camp, when nothing was moving and the possibility of smiling was not within reach, but I'm sure whatever toll our new existence had on me as a child was infinitely greater for him. I had to make sacrifices, but I didn't carry the burden of deciding between bad and worse options.

Whenever my father had a few extra shillings, he took me to the camp's market where he bought a Fanta and a Swiss roll that we'd split. He wanted to make sure I was okay, which I was, because I had him. As a child who knew how sad this world could be, every day I counted my blessings that I wasn't an orphan.

**EVENTUALLY WE ADAPTED TO OUR NEW LIVES AND CAME OUT** of our funk. Not only did the adults begin to dream and talk about what an exit from Utange might look like, but they also started to improve life in the camp in everything from do-it-yourself malaria prevention to card games.

I played a lot of cards with the men of the camp. Friends of my uncles and my father would stop in front of our tent to play *dabakaeri*, a Somali betting game similar to poker. Baba loved it. My oldest brother was really good. I was good, too, and thought nothing of jumping into their game—despite the fact that my dad didn't like the idea of his ten-year-old daughter shuffling a deck surrounded by a group of older men.

I have a great regard for older people but am not intimidated by them—which is why we always get along. I was raised to respect my elders and understand their value. At the same time, though, there was no hierarchy by age in my family. There was no contradiction in this for me, but it could appear that way to others. I always made sure an older guest was the first to sit down and be served at a meal, but I would also be the first one to cut them off if I disagreed with something they said. So it was at the card table; I fetched whatever the players needed but didn't let them win. (After we left Utange, I stopped playing cards, since the memory is a happy one that also makes me sad.)

Even if Aabe had wanted to hide the card games from me, he couldn't have. The camp was made up of row after row of beige tents that faced each other with little space between. The physical closeness meant that we were all in one another's business. Games were the least of it. It wasn't out of the ordinary to see and hear people having sex. Everything in Utange was out in the open.

So much so that you could find a bunch of things to entertain yourself with in the camp. My favorite activity was going to the "movies," which was just a television with bunny ears set up in a shed by an enterprising Kenyan in the neighboring town. For a few shillings, you could watch a show or movie airing on TV. That is, if you were able to sneak into town by crossing the barbed-wire fence. My friends and I used to squeeze through small tears in the chain-link fence. Sometimes, on our way back into the camp, we would get chased by locals, who didn't like it when refugees crossed into town. Unable to make it back to the gaps in the fence without getting caught, we had to climb over the barbed wire (I still have the scars).

Watching TV was worth the risk, though. Back in Somalia I had been addicted to watching movies—more specifically, Bollywood films. They were dramatic and full of joy. They also portrayed blended families, which normalized the kind of hardships and battles that naturally occur when different generations live together under one roof. In Utange, I was happy to watch anything at all that moved on the screen.

The kids in the camp found diversions among the day-to-day dreariness. We jumped rope, played soccer, hide-and-seek, and tag. Sometimes we made up scavenger hunts in which players had to find odd items throughout the camp. We also spent hours flicking bottle caps and matchboxes with the goal of getting them to land right side up.

No matter how long the games went on for, I returned periodically to my family's tent to see if I was needed. My role as "errand boy" started from early morning and went all day until we gathered at night under the starry sky to eat dinner and listen to music or the BBC on the radio. My first errand of the day was to fetch water for Baba, after I heard him read the morning prayer and sat awhile reading the Quran with him.

There were watering stations throughout the camp where people lined up with plastic jugs to fill when they finally arrived at large cement blocks with faucets. The water line was always a fight, which was why my family sent me. Not only because I was little, but because I was never going to let anyone cut me in line. Altercations broke out, because the water ran out at the stations. Then you had to go to another section of the camp for water. Sometimes I spent half the day going from station to station in search of water.

The other line known for its battles was the one for the bathroom. Standing in line for hours when you need a toilet is an obvious stressor. The water I collected from the water lines in the morning was used for everything—including washing our clothes, which was done by hand outside our tent, and washing ourselves, by taking water in a bucket to the bathroom and pouring it over yourself once you got to a stall.

I also stood in line for our food, such as rice, beans, flour, or oil. Various food items were scheduled for distribution throughout the week and month. Whenever families had an excess of anything, they traded or sold it to the locals. For our family, that was always kidney beans, which we traded for kerosene for our lamps, batteries for the radio, or good firewood, since we cooked over open campfires. If we didn't have firewood, I gathered kindling from the trees and bushes that ringed the camps—and that provided a fertile breeding ground for malaria-carrying mosquitoes.

Illness still posed a constant threat, but there were new dangers. As the refugee settlement grew in size since its inception, the nearby Kenyans began to resent our presence. Two years after the Office of the UN High Commissioner for Refugees and the Kenyan government set up the camps, the population had grown to 334,000 refugees. What did these sprawling squalid camps bring the local Kenyans? Other than black-market trading with the refugees, the only real economic benefits to the locals were the UN Refugee Agency contracts for building the camps. Because of the financial incentive in rebuilding, the camps were burned down several times.

**WE LIVED THROUGH TWO OR THREE FIRES AT THE CAMP DURING** our stay—each time, I smelled smoke and heard the loud cracking of the bamboo frames that held up the tents going up in flames. That was the signal to flee, because nobody knew where the fire was going to stop. We watched in terror as the ignition sparks jumped from one point to the next. A section of the camp would be on fire for a few seconds only to disappear and then reappear closer. While we grabbed whatever belongings we could carry—

a few pieces of clothing or a cooking pot—the heat became oppressive: a roaring fire in the noonday sun in a hot, dry land. Men, women, children, old people—all ran away from the jumping fire.

Once the fire was over, we all went right back to where our tent had been and started cleaning up. Sometimes people would try to claim another family's spot (the tents weren't numbered), and then the mess would begin. Weeks of chaos ensued as people fought over territory. But at other times people would come together during the rebuilding and help those who had lost the most get back on their feet. Because our tents were in Zone C and the fires always started in Zone A—it bordered the local villages—they never burned down. So we would often take in a displaced family or two from the other zones, who would stay in our tent for a while until they could rebuild.

Although we witnessed the worst of human nature in Utange, we also witnessed the best of it. The greatest lesson I came away with from my time in the refugee camp is that your today doesn't get to determine your tomorrow. Everything in life is fluid. Pride, strength, and responsibility—all of those notions are the domain of people in comfort and safety. When you're facing death, you're not guided by your importance or your past, and you certainly don't worry about whether your pride is intact.

Again and again, I witnessed that if you can push

through whatever is happening today, tomorrow might be worse, but it could also be better. The only option for the human spirit is to keep going.

And people in the Utange camp did. They buried loved ones in makeshift graves and then went to play soccer. Pain and death. Laughter and love. This is what it is. You just move on.

# 5

# AMERICAN DREAM

*1992–1995*
*Utange Refugee Camp,*
*Kenya, to New York, New York*

I sat with my grandfather outside our tent practicing for my Quran recital. A year after we arrived in the camp, I was good enough to compete with the boys. I was practically the only girl in the makeshift *dugsi*, which means Islamic school in Somali, that had been set up in a tent. Various Islamic teachers, all strict and authoritarian, had us recite long passages from memory for a couple of hours every day. Now that I was spending nearly every morning going over my lessons with him, Baba no longer had to yell at me. I was always ready with the recitation.

My dad arrived at the tent late in the afternoon. It was between the two prayers and an unusual time for Aabe to be back. He sat with my grandfather, and the two began talking. I didn't take any interest in their conversation until other family members started to huddle around them. Then I gathered close, too. That's when I heard what they were discussing: leaving.

"There is Norway or Canada," Baba said.

"Sweden is also a possibility," my father said.

"The United States is different," Baba interrupted. "Only in America can you ultimately become an American. Everywhere else we will always feel like a guest."

As soon as I understood what they were talking about, I got so mad.

Aabe had applied for our family to be resettled in another country. The process was a long one that started with Aabe registering us with the United Nations High Commissioner for Refugees (UNHCR)—the UN agency that decides whether a person meets the criteria of "refugee." Refugees are those fleeing war, persecution, or other violence in their home country, often in large groups and to countries that border their own.

If the UNHCR gives a person refugee status, the next step is finding a "durable solution" for their particular situation. There are only three options for refugees: returning to their native country, staying in the host country permanently (which in our case meant staying in Kenya), or resettling in another country permanently. The last option is the hardest to achieve for obvious reasons—when we were in Utange, resettlement to third countries like the United States was offered to only a small number of refugees.

There are extensive background checks, interviews, and medical screenings before a refugee can qualify for the limited number of spots to be resettled in another, stable country. But like the other two solutions, resettlement does not require a personal sponsor. If a country decides to take you in, the government is your sponsor.

The U.S. Refugee Admissions Program is a group of nonprofit organizations and federal agencies working both in the United States and abroad to figure out just who should be given a golden ticket to America. Every year that number can change based on the president and Congress, which determine the cap on refugees. In 1990, around 2,500 Somalis were living in the U.S., according to UN estimates. Two years later, on the heels of the humanitarian crises caused by the civil war, the United States began issuing new visas to Somali refugees. Not too many, though. According to the United States Immigration and Naturalization Service, 10,464 Somalis entered the country between 1986 and 1996.

As far as I was concerned, I didn't care if we never left. I had fully acclimated to my life in Utange. Sure, I had to wait in a long line for water every day and had few opportunities for any schooling other than my religious education, but I had friends and a routine. I had found normalcy and so had the adults, who weren't in shock anymore. The fact that Baba now had the mental space to help me with my Quran was a sign that we had fully settled in.

"I don't want to go anywhere!" I cried.

Everyone looked at me like I was crazy.

"What is *wrong* with you?" my father asked.

"Are we all going to go together?" I asked. "Because I don't want to go anywhere without you or Baba."

The journey out of Somalia, during which our separation had caused me great anxiety, felt as fresh as if it had happened yesterday.

Aabe tried to reassure me, but I barraged him with questions. How was this whole thing going to work, anyway? What about the other people in the camp? Where were they going?

When I tried to make the case for us to stay in the camp by stating that we would never see our friends again, my siblings started to make fun of me. They called me silly, which made me even angrier. But they didn't care, since they had stopped paying attention to me and had returned to the discussion of where we might move.

That was the first and last time I was included in a conversation about our sponsorship—a process that began with my father walking to the international refugee resettlement office outside the camp every day to look at a sheet hanging on its door with names of those listed to be interviewed.

A couple of months after I first heard about the possibility of us leaving Utange, he came back with the news that it was our turn.

**ON THE LONG, LONG LINE TO BE INTERVIEWED BY A UNHCR OF**ficial, Aabe, my siblings, and everyone else waiting with us started to panic. From what I understood, there was some lady named Pamela, and if your case was assigned to her, you were done. You weren't going anywhere except back to the camp. For hours, we watched people come out of the office with their fates written clearly across their faces. Some were happy, some frustrated, some in tears. None of them had received any kind of official notice of the result of their interview, but as one of my sisters said, "You don't need a piece of paper to know."

I found the interview unremarkable, perhaps because only applicants who were eighteen years old and older could be asked questions. So while my dad and four oldest siblings sat at a table facing a panel that included an interpreter, we three youngest sat off to the side, and I spaced out. Naturally they took all our names and birth dates. They said something about bedding, too. And then we were done.

It wasn't until maybe a year later that we learned that not only had we passed our initial screening, but my father, siblings, and I had earned one of those golden tickets to America. Aabe agreed with Baba that if we couldn't go home, the United States was our best option for making a new one.

NOW WE HAD TO GO THROUGH A HEALTH SCREENING, WHICH brought a fresh wave of panic. There were a host of physical conditions from tuberculosis to pregnancy that could cause one's application to be denied. There were horror stories where a whole family would have to leave a young woman behind just because she was pregnant.

Luckily, we all passed our health test and received the many vaccinations necessary to come to America. An organization called Catholic Charities arranged for us to live in a place called Arlington, in the state of Virginia. The next step was for us to complete a pre-departure orientation program in Nairobi for our new life in the United States.

Aabe, my siblings, and I—who were all part of the same visa process, since we are immediate family—traveled nearly three hundred miles to the capital early in the morning. The orientation program consisted of watching a video of images like sheaves of wheat blowing in the wind against the backdrop of a perfect blue sky (I assume this was a nod to "amber waves of grain," but I wouldn't learn "America the Beautiful" until much later) and rows of beautiful homes lined with white picket fences. One scene made a particularly strong impression on me. A family of four—father, mother, daughter, son—smiled at one another around a table with the biggest roasted chicken I had ever seen. (Now I know it was probably a turkey, in a depiction of Thanksgiving. But they didn't say that in the video.) The whole montage, from escalators and malls to fields of grain and that gigantic chicken, offered a picture of America as a land of abundance. To a bunch of refugees like us, this was something to get really excited about.

Our family was put up in a hotel room in Nairobi because of the long distance to the camp. It was cramped quarters, but no one wanted to leave, since stepping out on the city streets could mean destroying the whole process. In a classic bureaucratic catch-22, we had to come to Nairobi to get papers so we could go to America, but we didn't have the documentation we needed to travel outside the camp. Had we been arrested in Nairobi because of our lack of papers, then we *couldn't* go to America.

I cried and cried, asking to leave the cramped room and return to Utange to be with Baba, who was awaiting his own refugee status from the UNHCR. Finally my sister agreed to take the eight-hour bus trip back with me. "You are so annoying," Iskufilan said. "Who doesn't want to be in Nairobi and stay in a hotel? Who wants to go back to a camp?"

**WHEN THE FIRST GROUP OF FAMILIES SPONSORED FOR RESET**tlement boarded buses to take them to the airport, a large number of us from the camp gathered to say goodbye. There were hugs and tears, well-wishes and messages. It was a big deal. When our family left, it was a different story. Hardly anybody was there to say goodbye to my father, my six siblings, and me. A few of my siblings' friends came, and of course Baba, whom I was nervous about leaving behind, because we didn't know when or if he would be eligible for resettlement.

I didn't fully accept that we were leaving. When we left for the airport in Nairobi, somehow I convinced myself that we would be going back to Utange one more time before our final trip out of the country, even though on some level I knew that didn't make any sense. I wasn't interested in logic or reality. Every escape meant that I was leaving people behind and that my life was not going to be how it had been the day before. "Better" couldn't promise me happiness, but it would permanently alter my current norm. The anticipation of another unknown life gave me anxiety, and my coping mechanism was to disassociate. The discomfort of leaving made me disengage from it.

We boarded a plane that was filled with refugees, not

just from Somalia but also from Rwanda, Sudan, and Ethiopia. People were dressed in the strangest ways to go to America. A man handing his boarding pass to a flight attendant wore a suit that was at least two sizes too big for him. Two little girls, testing out the tables that popped out from the seats in front of them, were in Easter Sunday dresses as if they were about to attend a holiday party.

We ourselves weren't wearing tracksuits or comfortable casual clothes like most plane passengers wear, either. The outfits we had chosen to travel in, like those of our fellow passengers, were more in the style you see in old American movies when immigrants arrived from Europe by ship and disembarked wearing dresses, suits, and nice hats—and were met by hanky-waving crowds. The suits, dresses, shiny shoes, and hats we wore were a symbol of what we expected from *our* arrival in America.

There was an unspoken fantasy that when we came to America we would be greeted by its citizens, whom we needed to impress in order to fit in, so that we could land a good job, go to the right school, and move into one of those beautiful homes with the white picket fences we had seen in the orientation video. The refugees on our plane were dressed in such a way as to prove that we were going to rise to the occasion. We had been given a golden ticket, and for that we had to look like gold.

I'm not sure we achieved exactly that. While you could tell that the passengers were in their very best outfits, and that we had worked hard not to show up in America in rags, it looked like all of us were wearing other people's clothes—because we were. We had found jackets, blouses, and pants at a flea market in Eastleigh, Nairobi, clothes that didn't fully fit but that we could afford and that somehow signaled prosperity either in cut or material.

I was also wearing used clothes. To travel to America I wore peak nineties-style jeans with a button-fly high-waisted top and tapered legs. The coat I decided would protect me from North American winters was a burgundy Ultrasuede bomber jacket. My shiny shoes, intended for dance class, were so smooth on the bottom that I could slide across the airport floors while we killed time during an eight-hour layover in Amsterdam.

I don't know how, but Aabe had a little bit of money for the trip. He bought me a shortcake in the Amsterdam airport, because I was starving. Like many of the other refugees on our flight, I had slept for most of the trip and hadn't eaten. I had woken up once or twice, and each time my father wasn't in his seat. A smoker at the time, he had gone to the smoking cabin for a cigarette. I couldn't fall back to sleep until I saw him come down the stairs.

We circled the airport over Amsterdam for what seemed like forever. Too engrossed by the airport itself, I didn't even notice the other travelers. The excitement of being in this space filled with well-stocked shops, brightly colored magazines, TV screens, and music was too overwhelming. If this was what an airport in Holland was like, I couldn't imagine what awaited us in America.

**FINALLY WE LANDED IN NEW YORK, AND BECAUSE IT WAS TOO** late to continue on to our final destination of Arlington, Virginia, we planned to stay the night at a hotel in Manhattan. My father, clutching our IOM bag, ushered my siblings and me into a cab. The white plastic bag from the International Organization for Migration (IOM) is famous among refugees. It contains all-important documents from refugee identity cards to medical records, so people guard it more than they guard their own children.

Aabe was no exception, holding that bag tighter than any of us as we traveled in the middle of the night.

Through the window of the taxi I watched the darkened highways become city streets—and I was appalled by what I saw. Trash everywhere.

For what must have been ten minutes, I was glued to the window, taking in all the trash. While New York is known for being a particularly dirty and densely populated city all the time, at night, people put the trash from buildings out on the sidewalks for pickup in the early morning, so it gets a lot worse. Large pyramids, some even taller than I was, of black trash bags lined the streets—as if New Yorkers were preparing for a levee to break.

Perhaps if I had been walking on the street I could have looked up and appreciated the skyscrapers, but in the taxi all that was at my eye level was trash: garbage cans overflowing on every corner; food wrappers, newspapers, and plastic bags skipping down the streets.

I was distraught. Where am I? Where is the big, beautiful city? All I see is garbage. How . . . how is this America?

The images of America I had arrived with came primarily from the cowboy movies that we used to watch in Somalia. But I knew the scenes from Westerns—animals in barnyards, people riding horses, cowboys shooting each other outside saloons—were not from present-day America. The other source of information I had was the orientation video, where the America presented to us—the America we watched go by in pictures of immaculate suburbs, bountiful farms, and thriving malls—was perfect.

My twelve-year-old brain was freaking out, trying to reconcile the urban landscape before me with the video I had watched in orientation.

Every cosmopolitan area has its own challenges. I grew up in a city and then lived in a refugee camp only six miles from Mombasa. How did I expect there *not* to be trash? Or poverty? Or crime? But there's a breakdown for the refugee in the separation between what you are told and what is real. To be promised a utopia only to be brought to a city or town that might have a little less trash and crime and a few more buildings than where you came from is disorienting and disappointing.

Finally I turned to my father and said out loud what I was thinking: “This isn’t America.”

He looked down at me, worn with travel and concern. “Shh. You ask too many questions,” he said. “This isn’t our America. We’ll get to our America.”

# 6

# HELLO AND SHUT UP

*1995–1997*
*Arlington, Virginia*

From New York we flew to D.C. and then rode the bus to Arlington. It was my second day in the country, and I still hadn't found the America that I felt had been promised to me. In the light of early morning, I saw graffiti, more trash, and to my greatest horror, homeless people sleeping right there on the street.

When we lived in Somalia, in the big city, even next to the major outdoor market, I never saw a person who slept on the street while others just went about their day. That concept didn't exist in my country's communal society. For sure there were beggars. While we were eating, sometimes they would stop by our compound with kids in tow asking for money or food. They weren't described as homeless, because they had people to take care of them. As a kid, I don't remember ever seeing anyone sleeping on the street. Instead, Baba explained that these were people who were faced with challenging circumstances and didn't have what we had. Because of that, we had to help them where and when we could. And, at least in my mind as a child, they had somewhere to go at the end of the day.

In Kenya, we saw groups of young men hanging out on the street in dirty clothes. The adults talked about how they were on drugs, but they didn't seem completely dejected like the old woman I saw lying across a New York City park bench with only her shopping cart as protection. The Kenyan street kids weren't treated like objects to be walked around and ignored by residents and commuters making their way down into the subway.

In retrospect, when my father had said we hadn't yet made it to our America, I think he meant that New York City wasn't where we were staying, that we still needed to travel a little bit more to get to the city that we would call home. I was looking for an escape from devastation into something wonderful, the wonderful place I had been shown in the video, a reality that I almost thought couldn't be a reality. But we had sacrificed and invested so much in this journey that I couldn't accept inequality and suffering upon arrival—just a different kind. I was—and still am—in search of America as that more perfect place.

**IN ARLINGTON, ALL SEVEN OF US MOVED INTO A TWO-BEDROOM** apartment, applied for our Social Security cards, and went through another medical screening that included getting all the same immunizations we had just received in Kenya. My father tried to protest, worrying that the extra shots might make us sick, but if my sisters and brothers and I were going to school, we needed to have these medical records, so we took the shots.

It was March, and cold out. The clothing we'd brought from Africa wasn't warm enough, so we went to the Goodwill to get coats and other things. That was fun. I got two coats, but I hadn't figured out the shoe business yet. I was still in those shiny thin-soled dance shoes that were definitely not for winter. On the first day of school, when I got off the bus, I immediately slipped on the ice and fell hard on my butt. I hurt myself, and the other kids laughed. It was bad.

I don't remember getting even the most basic guidance from anyone—for example, that you need shoes with thick soles in the winter. We didn't have a church or a mosque or any other kind of community to help us with the minimum we needed to know as we transitioned to our new life. My dad picked up tidbits of information here and there, because he spoke a little English. Me? All I knew how to say in English when I arrived was *hello* and *shut up*.

As one might imagine, it wasn't easy navigating what Americans called middle school with only those two phrases.

My first trip to my new school was to take a placement test. It was in English, which I didn't know. But that was the point—to see what I didn't know.

I hadn't had any formal schooling for about four years. I'd attended the madrassa, which I enjoyed. Not long before we left, a school opened up in the second camp that we were in, a camp just for Benadiri, where my grandfather was one of the leaders. I didn't get to attend more than a few classes, however, before it was time for us to go.

Although I was in the fourth grade when war broke out in Somalia, because I had started school early for my age, I was placed in sixth grade from the results of my test. I got a quick walk-through of the school by an administrator who showed me the place where I was to be dropped off and picked up by the bus.

I was overly excited about going to school, because I hadn't been in such a long time. I looked forward to having a structure to my day and, as with America itself, had an idealized version of what my school experience was going to be.

EDUCATION HAD ALWAYS BEEN DEEPLY PRIZED ON BOTH MY mother's and father's sides of the family. For a period when Awoowe, my paternal grandfather, lived with us back in Somalia, he would call me into his room each morning before I left for school to make sure I understood that I was a beneficiary of the attributes of Araweelo, a Somali queen who fought for the rights of women and the disenfranchised.

"You should walk in the world as a proud person," he advised. He wanted to make his grandchildren strong and wise by embedding in us the history of our ancestors. His lessons were that our place in society is informed by where we came from, and external appearances are less important than how we feel inside about ourselves.

As I began school with only a couple of months left in the semester, I didn't fully consider the cultural differences between the school that I was nostalgic for and the one that I had arrived in.

My first week was very hard. When we got off the bus, most of the kids went into the cafeteria to wait for the first bell to ring. But I didn't realize I could go inside, too, because that hadn't been part of my walk-through instructions. I was told I could get off the bus here, then go to this classroom and that classroom. I had this locker with this number. No one had said anything about entering the cafeteria in the morning. Arlington in March isn't very cold, but it would drizzle and sometimes even snow lightly. I sat freezing on a bench outside, looking through the cafeteria windows at kids laughing and having breakfast. That had to be a special privilege. Because I was afraid of looking like a fool, my pride kept me from walking into a situation that I couldn't bullshit my way out of because I didn't speak English.

Because of my lack of English, I was put in the English Language Learner (ELL) program, for kids who needed extra help or attention either because English wasn't their first language or they had learning differences. In that class, I was surrounded by Latino students, who I thought were speaking English, but in retrospect were probably speaking Spanish. Whatever the language, they could at least talk to one another. I was the only one who was deafly moving around in space.

School, it turned out, sucked. It wasn't just that I couldn't understand anyone. I didn't even have a friendly face for moral support. My three oldest siblings, all over eighteen, attended a GED program where they could also work. My other brothers and sisters were in high school—together. They didn't struggle like I did. Not only were they older and had enough English to get by, but they also had one another. I hated that the three of them were going through the process together while I was essentially on my own in middle school. It didn't matter whether they could have helped me or not; just having them with me would have been comforting.

But I didn't, so I continued to sit in the rain, cold, and snow until one day a security guard, who must have watched me sit there for a few days, came over and led me inside. He brought me to the line for food and told me to stand in it. I followed the line and picked out what I now know to be French toast, which was delicious.

THAT WAS A MOMENT OF KINDNESS IN WHAT WAS OTHERWISE A really rough first couple of months. I was in fights constantly that started when anyone stared at me. I stared back, and if they said something, which of course I couldn't understand, I usually decided to hit them first, assuming they were going to hit me. I wasn't afraid, and I wanted people to know it.

I fought mostly boys and mostly white kids, or at least I thought they were white.

As an African, I had grown up a black kid in an all-black *everything.* I could probably count on one hand the number of non-black people I had met in my twelve years. And by *white*, I meant a lighter shade of the typical dark-complexioned Somali. So Arab Somalis like my grandfather didn't count. White people were the few Italians, holdovers from when Somalia was an Italian colony, or the one Canadian I encountered.

In the refugee camp, we saw many more whites—Americans, Germans, and Swedes—working for the Red Cross. I was never so confused as when I heard white people, descendants of British colonialists, in a Kenyan bank we walked through to get to our medical examination, speaking fluent Swahili.

While I knew about Native Americans and black people, whose ancestors were African slaves, my idea of America was that it was majority white. But my conception of white was very different from the American construct. There is no Somali translation for the word *Caucasian*. The word we use describes an actual skin tone, the way you appear. In that context and in my head, almost everybody in the school, including kids from the Dominican Republic and El Salvador, was white.

In the very beginning, I listened while the teacher read out attendance and heard some Somali-sounding names. But they turned out to be Eritrean and Ethiopian kids who also happened to be Muslim. I became fixated on two sisters, one in my grade and the other in eighth grade, who had actual Somali names. But that was the only Somali thing about them. They identified as black, a concept I had no understanding of at the time. I was so frustrated with them, because I understood that they were Somali even if they didn't. We didn't develop a relationship because I couldn't communicate with them. They were so close, yet so far away.

My anger and subsequent fighting landed me in detention for most of my sixth-grade year. And while I attribute my quickly picking up written English and getting good grades to all the hours I had to sit in a quiet room reading and writing, my getting in trouble so much was hard on my dad.

Aabe got calls from the principal's office all the time, which meant that he needed to leave work to come to school for a meeting or to pick me up. His first job was at the airport, helping people who needed assistance getting to their gates. Eventually, after passing the exam for taxi drivers, he started driving a cab.

After explaining to his boss at work why he had to leave yet *again*, Dad would show up at school so upset. Despite being really angry with me, he tried his best to explain to the principal how I interpreted the other students' actions as their sizing me up and thought I had to defend myself. I was lost in translation.

"Somebody looking at her doesn't mean they're going to hurt her," the principal told my dad. "Just because she misunderstands what's about to happen doesn't mean she has a right to fight people."

Fighting didn't feel like a choice. It was a part of me. Respect goes both ways. If I was showing up respectfully, I expected my peers to do the same. And that's far from what they were doing.

**THE EARLY DAYS IN MIDDLE SCHOOL WERE FILLED WITH TAUNTS** and worse. One day, someone pushed me from behind while I was walking down the stairs. Catching myself on the banister, I didn't fall. With my ego far more bruised than my body, I turned around and demanded to know who had pushed me. Kids pointed at a boy already ducking out of the stairway, but I chased him down the hallway.

I couldn't let that act of aggression go unchecked. Just as when I was in elementary school back in Somalia and a boy tried to bully another student into coming to the chalkboard while the teacher was out of the class, I could see a structure forming. Today he pushed me and I just stumbled. Tomorrow, though, he'd push me again and I would fall. And the next day? I had to set boundaries.

In the principal's office, where I was hauled in for fighting, the boy insisted he hadn't done anything. "It wasn't me!" he said.

"Somebody pushed me, and you were standing there," I retorted.

My dad, who had come to the principal's office from work, translated for them and then turned to me to say his own piece in Somali: "What is wrong with you?"

*Me?* I didn't push, pick fights, or bother anything. I just did my best to function within the confines of the instructions I had been given.

The language barrier was not the whole story. By the time I returned to school after summer vacation to start seventh grade, I was nearly fluent. Dad imposed a rigorous system at home where we all had to talk English to one another, no matter how painful it was. My teachers liked me. In my yearbook from the end of that year, my teacher wrote, "Ilhan in '96: 'Hello and shut up.' Ilhan in '97: 'Hi, my name's Ilhan. I want to be your friend.'" I had language to communicate with others and navigate life. Still, school remained hard. The problem was no longer one of miscommunication but reputation. I had become known as the kid who fights.

I certainly didn't do anything to change that perception. The littlest thing could set me off, as when a kid in my class started to tease me about a boy's having a crush on me. I was writing when he turned to another kid and said, "Johan likes Ilhan."

"Stop talking about that," I said.

"What are you going to do about it?"

I hit him, and then he hit me. While we were tumbling over chairs, he kept saying things that made me more and more angry until I got my hands around his neck. I pushed him up against the wall and wouldn't let go even after he began foaming from the mouth. I didn't release him until some teachers the kids had flagged down pulled me off of the boy. That was bad. I got sent home; I got in-school detention; I got after-school detention.

However, my worst fight was in eighth grade. It involved a clique of girls I had tussled with beginning in sixth grade. We all had PE together, and on this particular day, we were in the locker rooms changing after playing basketball. I wore leggings under my shorts, because as a Muslim, I didn't show bare legs or arms.

I don't know whether the girls knew that not having my body parts covered would be a lot worse than getting beat up or if they were just trying to sucker-punch me, but when I had my shirt halfway over my head I got hit.

With my shirt wrapped around my face, I lunged at anyone I could grab, trying to figure out who was around me. Even over the racket of banging metal locker doors, I could hear many girls shouting. Everyone always gathers for a fight.

Punches were coming at different parts of my body from different directions, so I knew there was more than one person actually hitting me. I wanted to rid myself of the shirt that was still half on, but I panicked that our coach, hearing the sounds of a brawl, would enter the locker room and see me in my sports bra.

Finally I got the shirt off and saw that I was surrounded by five girls, including a pair of sisters who lived on my block. Underneath me was Maria, a tiny girl from El Salvador. As I hit her, the others yelled, "She's pregnant!" I didn't stop hitting, even when they grabbed me by the hair on the back of my head and pulled so hard that whole clumps came out in their grip.

I wound up with a migraine from the hair they had ripped out, but if they expected me to submit—just like the rest of the kids who had stared at, bullied, or pushed me—they were dead wrong. I would hold my ground until it was completely understood I wasn't to be messed with.

This was about survival—I knew this from being a refugee. Without the protection of my siblings, a group of friends, or even people who looked and sounded like me, I was at the mercy of every bully and mean kid. My choices were clear. I could bleed every day and go cry in a corner. Or I could fight back and have people respect me.

# 7

# MINNESOTA NICE

*1997–2001*

*Arlington, Virginia, to Minneapolis, Minnesota*

School might have been hard, but everything else about my first couple of years in America was exciting. Going to the grocery store was always an experience. I was obsessed with canned tuna. One of my standbys was tuna sandwiches I made with ketchup. I can't eat that now, but at the time I was obsessed with them. Ketchup-tuna sandwiches and McDonald's fries.

Watching TV freely was another everyday luxury. I had spent four years going through a lot just to get a glimpse of a TV, and now it was a constant part of the day. I turned it on as soon as I got home from school. It was on when my family and I cooked and cleaned. We watched everything. People make fun of me when I talk about how I learned English from watching *Baywatch*. "They're not speaking English on *Baywatch*," a friend joked. I also experienced moments of extreme joy whenever I found what I perceived to be an American Bollywood–style movie with songs—what I now know are called musicals. A family favorite was *Cry-Baby* starring Johnny Depp. Whenever the John Waters film broke out in song, we would all scream and start singing, too.

We also walked a lot, mainly because, like with TV, we could. I begged everybody and anybody with a moment to spare to come walk with me. After years of confinement, walking was the ultimate freedom. It wasn't just the freedom of movement but also the freeing of your headspace that happens as you go. You're moving away from noise, and up until that point my whole life was noise—from a home filled with opinionated family members to the uncertainty of war to the chaos of the refugee camps.

I still love walking for the feeling of freedom it produces, which is why I never like to walk if I have a destination. However, if I'm roaming aimlessly, I can walk from Harlem to downtown, as I did the other night on a trip to New York City, without even realizing what I was doing.

When my sisters and brothers and I all lived together

our first year in Arlington, we used to go to the Pentagon City Mall, where I did dumb stuff like going up and down the escalator, just because I could. But the summer after sixth grade, all my siblings moved away. A cultural shift fed into their spirit. Having found jobs and opportunities for higher education outside of Virginia, they spread out across the country. This was very American.

That left just Aabe and me, and everything changed. Almost overnight, I went from someone with siblings, who were always in my business for good and for bad, to an only child. Of all the transitions I had been through—from upper middle class to refugee, African to American—perhaps this had the greatest impact. Because it was around this time that I began to display a level of willfulness my father wasn't necessarily comfortable with. He had raised me to be strong and independent—but within limits.

Or perhaps it was just that I was becoming a teenager. Either way, my dad decided to leave Arlington after eighth grade and move us to Minnesota. According to the 2000 Census, 11,164 Somalis lived in the state, the largest concentration in the country. More importantly, Baba was living there.

I was scared when we left Kenya that I would never see him again, but I was worried for nothing. Two months after we arrived in the United States we learned that the Utange camp was closing and the majority of its inhabitants, including Baba, had met the criteria for resettlement. When he first arrived in the States, he landed in Iowa, where he lived with a few of his sons and daughters. But as soon as he could, he relocated to Minneapolis. Baba was a cosmopolitan man. Not known for its cities, Iowa wasn't a great fit.

Arriving in Minneapolis, Aabe was just thankful he didn't have to be alone with me anymore.

**TWO MONTHS INTO MY FRESHMAN YEAR, I STARTED EDISON** High School. My classmates nicknamed me Virginia after a few of them asked where I was from. The name felt random and had appeared out of nowhere. Just like me.

Unlike my middle school, there were two or three dozen Somali students at Edison, which was exciting for me—almost overwhelming. After three years, overnight I was surrounded by people who understood things about my existence without my having to explain. There was a constant comfort that started with our common language. Even when we picked on one another, it was in ways that made sense to me.

During the school day, however, I wasn't with the majority of the Somali kids, since I was no longer in the ELL classes. Having decided back in Arlington that my education was the one element of my life I had full control over, and given the long hours of studying in detention, I had become a very good student.

Like Minneapolis itself, Edison's mainstream classes

were very diverse. Unfortunately, the differences among the student population proved more divisive than anything else. There were a lot of fights: everyone fought everyone. African Americans and African immigrants fought over who was blacker. Muslim kids and white ones fought over U.S. policy in the Middle East. Latinos against African Americans, Africans against Native Americans, and on and on.

Our lessons were chaotic. The bus was chaotic. Activities were chaotic. The funneling of students from all kinds of backgrounds and neighborhoods into one school building without any kind of preparation or process resulted in a volatile environment.

Unlike in middle school, where without the tools to communicate both linguistically and culturally I felt under attack, I no longer felt the need to get worked up if someone called me a name or to bicker in immature ways about stupid subjects.

With my fighting days behind me, the energy my peers were willing to expend on attacking one another seemed like such a huge waste.

I wasn't the only one who found the dysfunction in our school unproductive and unnecessary. I shared the majority view that it was hard enough just being a teenager. Adding a layer of racial politics was plain ridiculous.

Soon enough I became part of a group of students who had decided we wanted to do something to improve racial and cultural relations among the many different communities at Edison. With the idea of creating a coalition, we met with the principal and one of his administrators to walk through the kinds of challenges we were facing and the ways we could channel our energy into something positive. We were fortunate that they were willing to listen. (Many of us from that group still maintain relationships with the principal to this day.)

Unity in Diversity, the coalition we eventually formed, met in the lounge area our principal had offered for our organizing. We had come up with what was essentially a training program around diverse leadership. I don't know that any of us thought of ourselves as leaders at the time. Our inspiration was more pedestrian, something along the lines of students fighting on the bus so early in the morning that I wasn't even awake yet and they needed to calm down!

We ate lunch together and held retreats with freshmen, sophomores, juniors, and seniors. We recruited everyone with the express purpose of understanding the triggers of our racially charged environment and bridging the harmful divides. By the start of the next school year, the results of our labor were manifest in a much more peaceful high school experience, where if there was a fight, at least it was over a boy and not who was blacker.

**WHILE MY SCHOOL LIFE HAD CALMED DOWN AND HAD EVEN BE-**come fulfilling thanks to my growing role as a school leader, my home life conversely grew more difficult.

Living in Minneapolis was great for me, because I now had all these relatives I could visit. I liked to stay with Abti, one of my single uncles, or Baba, who lived by himself, in order to get away from my sister Shankaron, who mothered me.

Shankaron, who considered me her responsibility, didn't approve of my habits—like staying up until all hours of the night. I was like a vampire. I'm still like that. I come alive at night. That's why I especially liked staying with my grandfather, because we would stay up late watching basketball or random movies, old ones where he knew the actor who played the main character or whatever came on TNT.

Shankaron would tell on me, so I was constantly in trouble with my father.

"You know, Ilhan stays up all night to watch TV," she said. "I don't know how she gets good grades in school."

We fought all the time, because she didn't understand the typically American teenage tendencies I was developing. "Leave me alone; I want my space" has no place in the traditional Somali family, where it is normal for everybody to be in your business and for you to care what they thought. This conflict became a constant tension for my dad, who was caught between his belief in Somali respectability and in me as an individual.

In my early teens, Aabe had a list of demands for

me that started with basic appropriate behavior, like no drinking. Then he wanted me to get good grades. For that he offered a financial incentive. Every quarter he reviewed my report cards. If I got all As, I received three hundred dollars. For As and Bs, I'd get two hundred. If there was a single C, the amount went down to one hundred, and if I got a D in any subject, I got nothing.

The money I earned for my grades determined my budget for buying clothes, so I was very motivated. A hundred dollars was a pair of jeans and a shirt, which wasn't enough. My father also wasn't playing around. I couldn't get a C and then cry that a hundred dollars wasn't enough to buy much more than a pair of jeans and a shirt. A deal was a deal. So I mostly got straight As.

I understood and believed in my father's message, which was that only I could decide what I wanted to do in life, but that whatever it was, I should work my hardest to achieve it.

But then again, sometimes he seemed to reverse himself and he would punish me for things that had no bearing on my grades and didn't threaten any of the basic standards for morality he had set out for me from the time I was a child. "You told me that as long as I wasn't doing anything detrimental to my life success, you'd leave me alone," I argued.

So what if I stayed up late watching TV? Or dyed my hair?

When Beyoncé, who at the time was part of Destiny's

Child, dyed white-blond highlights around her face, I decided that was for me. In an attempt to be like Queen Bey, I, too, dyed strips of my hair blond, which on top of my very curly and very black hair was highly visible. Aabe called it the Lion's Mane.

My father had expected that I would fully absorb and adapt to life in the United States. I would go to school and participate in extracurricular activities. I would get a starter job. He envisioned me doing all the activities of an ordinary American teenager. But in his head, they would all be within the confines of what's culturally proper for a Muslim Somali, not the boundless flair of a pop star.

Through high school, Aabe developed an understanding of just what kind of child of his I was, how much of his trust and ability to parent I would test, and how alone he was in the process of raising me. My extended family, my dad's friends, the tight-knit Somali community—everybody tested my father's inherent liberalism. Would he stay true to the beliefs he and Baba had expressed when we still lived in Somalia, that as a female I had every right shared by my brothers, uncles, or any other male? Or could he be pressured into becoming a stereotypical father, imposing limitations on me because of my gender?

Much to my disappointment, people exploited his sense of pride and manliness in trying to manipulate him to curtail what they considered my bad behavior. "What Ilhan wears!" one relative complained. "You let her come home by herself so late," another judged.

I wasn't wearing anything more risqué than a tank top or doing anything more wild than going to the movies or a normal house party with kids from my school.

One time, somebody told him that I had been seen kissing a guy on the stairs, which sent him into a tirade. I fought back.

"I would never! Other people might do that, but I would never," I said.

"I raised you to be better than a slut!" he shouted.

As soon as the last word came out of his mouth, I saw he had shocked himself. While he would be just as mad if his son was caught kissing someone, he would never call him a slut.

Sometimes when we were fighting he did this—catch himself mid-yell—as if he was having a back-and-forth with himself. The expectations of himself as a man and as a father showed up in conflict. And he hated it, because he remembered the human who wanted to raise another human fully capable of making her own decisions, with total autonomy of body, mind, and spirit.

My father always exhorted me to "Go live your life." But he jokingly warned, "Don't do anything that would make me unable to sit with my peers, or anything that would make me throw you and myself in the Mississippi."

He never said what the "anything" I might do to cause those things was. But I had a few ideas.

I knew that being drunk was on that list. If someone caught me drunk, then he wouldn't be able to sit with them. If only he knew about my drunkenness, we were headed for the Mississippi.

Although most other kids my age went to clubs, I never went—not because I would be tempted to drink. No, I worried about there being a shooting or a fire and getting stuck or dying inside and my father's having to live with the fact that I had been in a club. My friends joked that I was more afraid of my father than I was afraid of God. "God is forgiving," I would say. "My father isn't."

My father raised me with a strict morality, not strict rules. He never said I needed to be home by a certain time, but he would have been angry if I came home late because I was at a party with my friends.

He never once said to me, "You can't leave the house unless you wear a hijab." A lot of the Muslim girls I went to school with would change their clothing as soon as they were out of their parents' house—the skirts got shorter as they got closer to school. This was a time when jeans with cutouts were popular. I told my dad I wanted a pair, and he went out and bought them for me. When I left the house, my friends, who put their cutout jeans on only when they were out of sight of their parents, were shocked. "You're the devil's child," one said. "How do you have a father who is okay with it?"

My dad was more concerned with the spirit than with the letter of the law. A revealing outfit upset him not for the body parts it revealed, but for the lack of self-esteem it demonstrated. He wouldn't have been fixated on the fact that a shirt didn't cover up my chest enough but rather on the fact that he had failed to teach me that my value as a human wasn't based on the extent of my body that I had on display. He would have blamed himself for not giving me enough self-esteem, resources, or love. The same would have been true if I were to have premarital sex, which I wouldn't have done anyway. But I didn't understand my father's prohibition on dating. I was guided by boundaries that I set for myself, not ones set by societal norms. Since I was tough on my own reasoning, I felt justified to interact in whatever way I wanted to.

**AHMED AND I FIRST MET WHEN I WAS SIXTEEN, AT A SPORTS** tournament that was part of a week of events in celebration of Somali independence. Earlier that day, I was getting my hair braided because I was going to be a bridesmaid in a wedding the following day. Traditionally, braids on a woman signify that she is unmarried. Once they are married, women cover their hair with a wrap. This was a nod to Somali nomadic culture that predated the country's Islamization, so the wrap was not a hijab but a smaller turban. There's a ceremony dedicated to this transformation, seven days after the wedding, called *Shaash Saar,* which means the putting on of a scarf.

The tradition came back into vogue after the civil war. When I was little, nobody did hair braiding for their wedding, but in the diaspora, people were looking for traditions that connected them to their African roots.

Whatever it was, I hated getting my hair braided. I hate people touching my hair. I asked to be the first one to get my hair braided so I could get it over with. The process, which involved a ton of hair extensions to make the long braids, made so much noise and hurt terribly. When I was done, I was so mad I didn't want to sit around and watch other people get their hair braided. I shouldn't have been somebody's bridesmaid; this whole girlie practice was not my scene.

I walked outside the beauty parlor with a wet towel wrapped around my head to soften my hair and because I hated the smell of burnt hair where they had cauterized the bottoms so that the braids wouldn't come undone. I was in a real mood. One of my friends, Faiza, said, "You've got to just go."

"I don't know where to go," I said.

Her boyfriend, who had stopped by to bring her money for her hair, was going to the games.

"Just go with him," she said.

Her boyfriend, Bile, happened to be one of my favorite humans, so I went along with him to a soccer game at Roosevelt High School. The parking lot was crowded with cars and people. Bile was really popular, and everyone wanted to talk to him. With the towel still draped over my head, I didn't feel like talking to anyone, so I walked straight over to the bleachers to watch the game.

From the bleachers I saw a young Somali guy arrive and approach Bile. As they talked, people huddled around them as if there was something interesting about the newcomer. Then they both approached the bleachers where I was sitting. I didn't catch his name when we were first introduced, but it didn't matter. He must have thought I was a crazy person with a towel on my head. "Lift the towel from your face," my friend's boyfriend said. But I made myself seem even crazier by answering no.

The game was cut short anyway, because a fight broke out. Now I was really annoyed. Some girls were fighting over a boy, which meant that our car was barricaded in. "I need to leave," I said, standing up. "My father will kill me if I'm involved in this . . ."

"Well, where does she want to go?" the mystery man said to Bile. "I could take her?"

Because he had arrived late, his car was way beyond the melee.

As he drove me back to the salon, he told me his name again: Ahmed. He had just flown in from Houston that morning to coach in the basketball tournament, but they didn't have a game that day, so he had decided to watch soccer. It was his first time in Minnesota, and he'd heard a lot about this gigantic mall.

"The Mall of America," I said from under my towel. "I can take you there."

For the rest of his stay in Minneapolis I acted as

Ahmed's unofficial tour guide, though his requests were often absurd. "Take me to Chicago Avenue," he asked.

"Chicago Avenue and what?"

"What do you mean?"

"That it's a long avenue."

We became really good friends during his trip and continued to be friends after he went home. I don't really know how or when the relationship transitioned into something more, but eventually it did.

Ahmed was me if I had been born a boy. Not only had he come to America from Somalia around the same age as I had and had experienced similar challenges, but we also shared a common spirit. He was very passionate about a lot of things and it showed. Whatever discussions he engaged in, he showed compassion and serious concern for the issues at hand. At other moments, he was a super-smiley fun human. He toggled between a happy-go-lucky attitude and an "our world will end unless we save it" approach. Either way, Ahmed was always brimming with life.

Ahmed transferred to college in Minneapolis, and our bond grew more serious. But our relationship wasn't just our own; it was a family affair.

My interaction with any boys, even friends, was an issue even before Ahmed came on the scene. While I was walking home from a movie with a group of friends, a guy put his hand on my shoulder. The physical gesture was totally innocent, but it sent me into a panic. My friends couldn't understand what was wrong.

"I have all of these family members who drive taxis," I explained. "I can't have one of them seeing me coming out of a movie with a boy's arm around me. They'll drive straight to my house and tell my father."

*Don't do anything that would make me unable to sit with my peers, or anything that would make me throw you and myself in the Mississippi.*

I'm not sure holding hands with a boy rose to the level of my father's words that ran in a loop in my head, but it would most certainly become a source of obsession among my siblings. I felt like everyone was always watching me, waiting for me to slip up. (My adolescence bears a lot of similarities to my life as a public figure. When people ask, "How do you feel living in a fishbowl?" I say, "I grew up like this. Sure, now there are cameras. But there were always eyes.")

I didn't try to hide from my family the fact that Ahmed was a boy who was more than a friend. Ahmed was a DJ, and when he moved to Minneapolis, I would go to his gigs. We were inseparable and in love.

It didn't make any sense to my father that I had a boyfriend—a cultural concept that I didn't question. I took my family's prohibition against dating for granted. For me, it was as culturally normal as people saying "Hello" when they greet someone. I also put it within the larger context in which I was growing up. Dating was a hassle for all teenagers whether they were Muslim or Somali or not. All of my friends were struggling with what their parents thought about the boyfriends or girlfriends they brought home.

Although I didn't feel my situation was unique, in my dad's opinion, you didn't date at all. You found someone to marry. And I was definitely too young for that in his opinion. But he didn't throw me in the Mississippi. Even though my relationship was a constant topic of conversation among my siblings at home. They didn't disapprove of Ahmed. On the contrary, they liked him a lot. By bringing the subject up all the time, however, they were incessantly reminding Aabe that his baby had a boyfriend—and that he wasn't doing anything about it. Thrust back into that irritating conflict of trusting me to do the right thing but feeling pressure to act like a stereotypical patriarch, he would get all worked up.

"I believe you would not do anything forbidden," Aabe told me. "But to say you're in love. You have no idea what that is."

# 8

# SETTLING DOWN

*2001–2003*
*Minneapolis, Minnesota*

The concept of marriage was nothing more to me than just that—a concept. And it wasn't exactly a natural concept, either. The idea of two people confining themselves in a legal and social construct because they entered into the metaphysical state of love always struck me as bizarre. The institution of marriage had everything to do with respectability and acceptance and not with a couple's devotion. I looked upon it as a performance that added nothing meaningful to what Ahmed and I were to each other.

Eventually, though, there was nowhere else for us to go if we wanted to stay together. The formalization of our relationship was about being accepted by God first, and then the acceptance of people followed. Only then could we create a proper and respectful life. Because dating wasn't accepted in our culture, to our families we were never boyfriend and girlfriend but rather people who were interested in someday getting married.

And someday, according to everyone around us, had finally come. We had been around each other enough, the families reasoned. It was time to take it to the next stage. So in 2001, before I was nineteen years old, the decision was made by our families that Ahmed and I would get married.

My dad was perhaps the only one who wasn't so eager for me to get married. He had expected me to go to college, earn a degree, find a profession, and become fully independent—and then maybe I could think about marriage. On the other hand, having his daughter in a relationship was an inconvenience to his conscience. He would have preferred his daughter have a legitimate reason in the eyes of God to hang out with a member of the opposite sex. This conflict tortured my father, who on some days thought our getting married was a great idea and on others, a horrible one. He finally resigned himself to the belief that Ahmed and I could build the kind of educated and employed life he had always envisioned for me, but together.

What did Ahmed and I think about all of this? Our love for each other was obvious. Other than that, we were just bystanders to the conversations that everybody else was having. The pressure became unbearable as our marriage turned into a constant source of conversation whenever we were with our families.

In August, the summer after I graduated from high school, Ahmed's family decided to ask for my hand in marriage. This is a formal tradition in which the males of the groom's family arrive to make the ask to the males related to the bride. I wasn't there, just as I wasn't there for the religious ceremony to marry Ahmed and me, a ceremony traditionally attended only by the male members of the family. My dad appeared on my behalf, accepting my willingness to enter into marriage. While cultural traditions in Islam are as varied as the places it spans, religiously a first-time bride is obligated to have a *wali*, a surrogate, stand in for her. The male can be present himself.

Our transition to married life was as seamless as our wedding ceremony. Ahmed and I were still young, and his family had moved to Minnesota, so it was natural for us to be around our families as much as we wanted to. We rented a one-bedroom apartment a few blocks from Aabe. Everybody on my side of the family loved Ahmed, particularly Baba. Ahmed, a big basketball fan, was a perfect companion to watch games with my grandfather.

Not much changed in our lives. We continued to

work and go to school just as we had before. The only difference was now we had unfettered access to be in each other's company privately.

Ahmed and I wanted to have a baby, so my goal was to find a college program where I could take classes, work, and raise a child. I did well in high school. Because of the summer classes my dad enrolled me in and my tutoring of ELL students in math and English, I had enough credits—more than enough credits—to graduate by the time I was a junior. I was unable to graduate early, since there were two classes—economics and civics—that only seniors could take and that I needed to earn my diploma.

With my good grades, extra credits, and the leadership role I had assumed in school in the diversity group and as a tutor, I had many options when it came to college. But I didn't know that. As was typical of many refugees, immigrants, or children who are the first to attend college in their family, I was totally ignorant of the vast higher education landscape and its wide variety not just in style but also in quality.

I found my school on TV. I had been watching when a commercial came on for a two-year associate's program that students could attend one day a week, one night, whatever worked for them. And it was right in the center of the Mall of America! Not only could I take classes on a flexible schedule, but the school was located in a major transportation hub, so I could drive there or take the bus. It seemed perfect.

Enrolling was just as smooth an experience as the school's advertisements promised. There was an entire administrative department of very helpful people devoted to finding me all the loans I was eligible for. They even took care of filling out the forms, so all I had to do was sign papers. Before I knew it I had a full financial aid package and was enrolled in a two-year program in accounting.

I had barely started school when on September 11, 2001, my classes were interrupted. Our country was under attack, and the Mall of America, the country's largest mall and the location of my school, could easily be a target. All the students and everyone else in the shopping center needed to evacuate immediately. I didn't know what was happening until I got home. Even then, it took a minute to register. My dad, sitting in front of the television, shouted for me to come look. Watching images of the Twin Towers ablaze and then falling, killing thousands of innocent people, I was upset and confused. My dad, in shock, was parroting every word the reporters said—as if it helped him register the horror unfolding before our eyes.

We eventually transitioned out of shock and into mourning. As Muslim Americans, however, we also had to contend with fear as well. We were held responsible, as a group, for the terrorist attacks or seen as a threat. Suddenly our religion was dangerous and our American-ness called into question. The comedian Hasan Minhaj put it so well in his Netflix special, *Homecoming King*: On

September 11, "everyone in America felt like their country was under attack," he said, but on September 12 and "so many nights after that, I felt like my family's love and loyalty for this country was under attack."

**A LITTLE MORE THAN A YEAR AFTER AHMED AND I WED, WE WERE** excited to learn that I was pregnant. Although we didn't talk a lot about why, we both wanted a child and thought having one was an important thing to do.

Pregnancy modified my personality. While I had never been shy, I became annoyingly extroverted. I normally don't really get too excited about things, but I was enamored with the idea that something—or someone—was growing inside of me. Obsessed with every detail, I spent a tremendous amount of time in the baby section of every place from the library to Target. I also asked everybody about everything I was going through.

When I say everybody, I mean *everybody.* At my jobs—at the preschool where I was a substitute teacher and the bank where I was in the debt recovery department—I needled my female coworkers about what foods they craved during their pregnancies, the strangest physical symptoms they experienced, and the items they brought with them to the hospital. I was the weird pregnant person who couldn't stop talking to the cashier about childbirth.

All my family members and a lot of my friends arranged their schedules so that they could spend time with me. "Ilhan can't be alone for five seconds or she's going to lose her mind," my friend Abdi said. During this outgoing pregnancy, I wanted to be around everyone, so that was fine by me.

My family enjoyed it for a while, because as a teenager out on the town I hadn't had time for them. Then they got sick of me and my one-track mind. They tried to ignore my phone calls. When I was around them, they tried to change the subject. It got to the point where Shankaron, now a nurse teaching a class to prospective doulas, recruited one of her students to earn her practice hours working with me. "You have somebody you can annoy," Shankaron said. "Please just stop talking to us about this baby."

I didn't just want to talk about my baby all the time; I also wanted to see it as much as possible. I made many trips to the hospital, armed each time with a new excuse as to why I needed an ultrasound. The medical staff cautioned me about adverse effects to the fetus from multiple ultrasounds, but I was undeterred.

Two weeks past my due date I woke up to discover my water had broken. We went to the hospital, where I learned I was three centimeters dilated. I could stay if I wanted to or come back. I chose the latter since, even though there was a huge blizzard that day, we lived nearby the hospital.

Back at home, my doula came to be by my side along with basically every single woman I'm related to—sisters, aunties, my mother-in-law. Everybody was hanging out as I went from the bed to the living room, doing my breathing exercises while the doula timed my contractions. I took a bath. We got pizza. At some point I made tea. Everyone was in agreement; I was fine.

Then around eight P.M. I randomly started to throw up. "Back home, women who didn't want the others to know they were in labor started throwing up when they couldn't hide it any longer," said Hawa, one of my aunts. "You could be very close now."

Having heard nightmarish stories about the pain of childbirth, I insisted, "I'm fine!"

But everyone, including the doula, insisted I go to the hospital—and they were right. By the time I got out of the car, I couldn't sit down in the wheelchair they wanted to transport me in. It turned out I was nine and a half centimeters dilated. There was not even time for an epidural.

In general, I have a high tolerance for pain. I'm not someone who takes medication. The contractions I had undergone so far felt nothing like what I had prepared myself for. It wasn't until about a half hour before I gave birth that I experienced real pain, and then a little after ten o'clock, our baby was born: a girl.

Despite my desire to know everything about the baby growing inside of me, Ahmed and I had decided to hold off learning the sex of our child. For some reason, though, we were sure it was a boy and had picked out a name in honor of one of Ahmed's friends, Adnan, who had come in from Texas a few weeks earlier to be present at the birth of his namesake. Our daughter, however, was born the day after he left. "She didn't want to break my heart," he said. "That's why she came after I left."

We called her Isra, the title of one of the surahs in the Quran that describes a journey the Prophet (peace be upon him) takes at night. Because she took this long, pleasant journey in arriving, we decided the name was a good one for her.

**IF CHILDBIRTH HAD BEEN SWIFT AND RELATIVELY PAINLESS, BE**coming a mother was anything but. None of the books I had read or advice I had received had prepared me for caring for a newborn. For example, I'm a caffeine addict. I drink lots of coffee and tea all day long, and none of it is decaf. My doctor said I needed to cut back on the caffeine when I breastfed or I would have a hyper infant who didn't sleep. "That's okay," I said, "because I don't really sleep, either."

What happened, though, was the exact opposite. I had a child who could not wake up. It was impossible to get Isra to drink all the milk she needed in the sole five seconds she was able to remain awake. I marched my sleepy baby back to the hospital to confront the doctor. "You told me this kid would be awake," I said. "Now fix her!"

A sleeping baby is supposedly every new parent's

dream, but not when your breasts are engorged with milk. Pumping bothered me for many reasons, including the terrible sucking and wheezing noise it made. I got mastitis and suffered from extremely high fevers almost daily, but I didn't want to take medication because I was breastfeeding. It was very torturous for the first weeks, as it is for so many new mothers.

Then it passed. Isra got bigger, and I realized how lucky I was to have a very pleasant child. She was as perfect as a baby as she is now as a girl. A good eater, she gained weight really fast and continued to sleep solidly.

The only aspect that worried me was that she seemed delayed in learning to walk (although in typical Isra fashion, she achieved the milestone the day after her first birthday as if that's the date she had decided upon).

Isra's birth was right around the time the United States invaded Iraq, and I had the news playing constantly on the television. When Isra took to crawling on her belly commando-style, Baba joked it was all the war coverage I was watching. And when I complained that Isra wasn't yet walking when my friends' babies of the same age already were, Baba said, "Just turn the TV off."

The truth is Baba, Aabe, and the rest of my family members—we were all glued to the television. To me it was clear that this was an illegal war, something that would be proved true in the fullness of time when the threat of weapons of mass destruction used as vindication for our invasion proved unfounded. Even though the people of Iraq were oppressed by a long and brutal dictatorship, war was only going to bring more destruction and devastation. As a child, I had experienced firsthand the ravages of war. And as an adult, I knew that what follows is usually decades of turmoil before anything close to normalcy is restored. During that time, more lives and futures are lost. War doesn't restore. It just robs. It takes everything.

# 9

# BLESSINGS

*2004–2005*

*Minneapolis, Minnesota, and Gothenburg, Sweden*

I walked into my auntie Fay's house and immediately crumpled on the couch. Isra soon came toddling in. With my face in my hands, I couldn't see her, but I felt my daughter, who wanted to be held, tugging on my arms.

"What did the doctor say?" Fay asked. She had watched Isra while I had my OB-GYN appointment.

"I'm having twins," I replied.

She gave me a look and then quickly said, "Children come with their own blessing."

"I don't know what blessing I'm supposed to feel at the moment. Children also come with more energy and time needed from me. And I don't have enough of either as it is."

When I found out I was pregnant again in the early summer of 2004, all I could think was "Why me?" A working mother, I was completely exhausted and overworked. I was also trying to finish school.

The two-year program I had picked turned out to be a lot harder to complete than I had understood when I enrolled. And a lot more expensive. When I was signing all the papers for my various financial aid packages that covered tuition, I didn't think about the other expenses I would have while pursuing my education—namely the cost of living and raising a family.

By the time I figured out that the financial aid I had received wasn't enough, it was too late. The problem wasn't just that I was being stretched too thin in trying to work, take care of my child, and complete my coursework, but I found out that the school wasn't even fully accredited. That meant that the individual classes I took would not transfer if I were to try to apply them to a traditional institution of higher learning like the University of Minnesota. You could transfer only a completed associate's degree. If I left after a year, all the time and money I had put into my degree so far would be wasted. Therefore, there was nothing to do but keep going.

I talked to the other students—most of them working adults with families like me—and we all felt similar anxieties and pressure. The school was too expensive for us to attend, but we couldn't afford to waste the money we had already spent on it. There didn't seem to be any good option.

The selling point of schools like the one I attended is easy access to registering and flexibility once you're enrolled. At the time, when administrators had eagerly walked me through the whole process, the opportunity seemed too great to pass up. I didn't know the questions to ask during the complex college admissions process that assumes everyone has a support system to guide them in making the best decision.

As a refugee with a single parent, I didn't have the traditional family who read college guides, toured campuses, helped with applications, and finally reviewed financial aid and loan applications. I was on my own like the other immigrants, veterans, and first people in their family to attend college—all of whom made up the bulk of students at my school. Overwhelmed by where to start in getting an affordable degree, we were happy to take any help offered to us. Unfortunately, at the same time that the school urged me to take out loans to pay for its fees, it didn't make sufficiently clear that, as I later discovered, many of its classes were unaccredited and could not be transferred to other schools. In retrospect, it feels very exploitative. My fellow students and I paid dearly for a school that filled out all the forms for us.

Part of the work I do now as a member of Congress is try to make sure that students get what they've been promised or what they can reasonably expect when they enroll in a school. That means very different things to different people, from job opportunities when you graduate from a trades program to being able to take your credits for an Accounting 101 class with you if you decide to transfer.

As a twenty-year-old mom, however, I had only one mission. That was to get my two-year associate's degree in business administration no matter what. I had changed my major from accounting early on and now was just going through the motions to get done by any means necessary.

The toll was visible. I was completely underweight. When I got pregnant for the second time I weighed maybe 89 pounds and was anemic. In pushing myself to graduate, I also pushed my body to extremes—not sleeping, not eating enough.

It was in this state that I learned I was carrying twins. On top of my schoolwork and workload, the prospect of two babies when I already had a child who had only just stopped breastfeeding and learned how to walk was beyond daunting. A blessing? The unexpected pregnancy was more like a test that I didn't know I needed.

I've gone through life with the understanding that every challenge comes to teach us something. But I couldn't imagine what the lesson was here. My life was closing in on me. I could hardly handle what I had going on; adding twins was madness.

They say God answers prayers, but the news of my

pregnancy seemed like the exact opposite action by our Creator. I felt like God was trying to mess with me. I felt like it was supposed to break me, and it worked. I felt quite broken.

But God's trials for me had only just begun. About thirteen weeks into the pregnancy, Ahmed and I were at the movies when I started bleeding. We went to the hospital, where I miscarried. When I cried, my dad called me "the best actress," because I had complained so much about being pregnant over the past three months. In that time, though, I had gotten over the shock and was starting to contemplate the mysteries of God's blessings. Now I had a whole new shock to absorb.

The trauma to my body from the miscarriage was exacerbated by the way my days were structured. Much like many other women, I was unable to put myself first. I was so preoccupied with making sure Isra was fed, my schoolwork was done, Baba had his meal, and I had checked in on this aunt or that sister that sometimes I didn't eat for days. Overextended, I kept going by drinking coffee. I went without food and sleep until I crashed and was hospitalized for dehydration. I had to be forced to drink and eat. My thyroid was also hyperactive, so I was running on energy that I didn't really have.

**WHEN I GOT PREGNANT A THIRD TIME, IN EARLY 2005, I NEEDED** to take a break. Despite the tumultuous events of my personal life, I had achieved my intended goal and earned my two-year degree. Whether it was firsthand knowledge of what it means to have a miscarriage or I was just tired, I had a completely different personality this time than I did during my first pregnancy. For the first time in my life, I became an introvert.

During that period, I was not a fan of humans. It was so extreme that noise of any sort bothered me. I had planned to take Isra to England for a couple of months, but the outskirts of London, where we were staying with family, was way too loud and crowded for me. I could handle it for only about a week before I decided we needed to move on.

When I announced we would be visiting other relatives in Sweden, my family back home freaked out. That's because my mother's first cousins, who lived in the small city of Gothenburg, are extremely religious, and at the time I wouldn't have been considered religious. Once in a while I would pray and fast, but nothing with any regularity. And I didn't cover my hair. What made my family back home the most anxious about my visit was the fact that I have no problem arguing with people about religion. They hated that I loved to challenge people on their beliefs.

"They don't shake hands," Baba said about the men before he rattled off a list of tips to keep me from shaming our family name.

I pictured all my aunties and uncles sitting around

talking about how the worst possible person was headed to Sweden. I know they worried that when someone told me he couldn't shake my hand, I might start a debate on religious interpretation. And they weren't totally off base. That's the kind of thing I was known to do.

Before I even stepped off the train in Gothenburg, I had committed a host of cultural taboos. The very fact that I was traveling by myself, not with my husband, was frowned upon by orthodox Muslims like my local relatives. On top of that, I wore jeans and a T-shirt that was stretched tightly over my very pregnant belly, displaying a conspicuous lack of modesty considered disrespectful in a conservative community. Traveling with a two-year-old, however, I was going to wear comfortable clothes no matter what anybody thought.

The final devilish touch was my uncovered hair with its Beyoncé-inspired highlights.

On the platform, I saw my mom's cousin walking toward me. Although we had never met before, I recognized him immediately, as he was the spitting image of one of my mother's younger brothers. Instinctively I rushed to give him the biggest hug in the world.

He froze. He was the imam at the local mosque and was probably thinking, *Somebody did not get the memo.* He didn't hug me back, but the earth didn't open and swallow me up, either. Instead, he chatted with me and took us back to the house, where I ended up staying for a long period of time.

The pious home in a quiet city was the restful and therapeutic environment I needed. My family in Minnesota had worried for nothing because I was far from the outgoing and argumentative person they knew and feared would offend our relatives. I was reserved during all my interactions with my mother's cousins, whom I found curious. Our conversations were vibrant and pleasant, but they all revolved around day-to-day subjects such as cooking or an outing to a park.

No one asked me any questions about our family back home. They didn't ask, "How's your grandfather doing?" Or inquire after my father or siblings. During the second week of my stay, extended family members arrived and we took chartered buses to a camp. Not a single one of the forty people wanted to know anything about relatives they had not seen since the war broke out. Did they even want to know if their family members in the United States were alive?

I pulled aside Aliyah, a cousin who had been really close with my mother, and asked her about this perceived indifference.

"People misunderstand us," she said, "but we essentially have only two rules. The first is that we worship to the maximum of what we understand is required of us." That meant no gossip or complaining to the point where if they asked me about our family members they could be inviting me to say bad things about them. I was more than welcome to talk about Baba's health or what my husband was studying. However, they weren't going to take responsibility for soliciting the information and risk breaking their second rule: "You can't truly be religious if you wrong others in your observance." So even if my cousin never talked about other people, if she asked me to, it was no better than her gossiping herself.

"Anything that is a transgression against you takes away from the good deed we wanted to fulfill," she said, "including pushing our beliefs. That, too, is a transgression that minimizes our worship."

It was why, she explained, nobody woke me up for prayer in the morning—something I had also wondered about and been a bit annoyed by.

"Imagine if we woke you up and the little one woke up, too," she said. "You might become stressed and internally curse us. Now we are going to pray, ask God to forgive us and to reward us. Why would we give it away to cursing by entering a place that was not our concern?"

She went on, "If God wanted you to wake up for prayer, he'd wake you up. And if you don't, that's your answer to God. Our letting you sleep is the good deed, because when God judges us, we're not going to be asked about *your* prayer. We're going to be asked about how we cared for you."

*Shit!* I thought. *I've spent my life around fake people of faith.*

Up until that moment, my main experience with religious people was their trying to give me a hard time about why I wasn't more religious.

Aliyah opened my mind to a radically different concept of what it meant to be devout. I discovered a solidly internal definition that rested on the care of one's own spiritual well-being and nothing else. Commenting that someone didn't pray enough or was dressing immodestly was born out of insecurity. A healthy religious practice is for you and you alone.

**ADNAN WAS BORN IN THE FALL OF 2005. AHMED'S FRIEND FI**nally got his namesake after what was my most agonizing labor. Because my water broke, I had to be induced. I don't know if it was the Pitocin the doctor used or Adnan himself, who was so active in my uterus that I barely slept during the pregnancy, but I experienced eye-popping pain. The threshold I had boasted about with Isra was seriously tested.

The practice of wearing the hijab, which I took on after Adnan's birth, seemed like the epitome of an external religious practice. It's why I had always struggled with it before I went to Sweden. I resented the fact that I had to cover myself for others. But after I returned to the States, I saw the hijab in a different context. It was no longer about what I'm supposed to be for *them* but what it does for me.

I fully got it. The hijab wasn't about a piece of cloth or the battle against objectification. Instead it was really a symbol of the purity of my presence in the world. It makes sense to me that I need to cover pieces of myself to preserve who I am and feel whole. I'm centered by the hijab, because it connects me to a whole set of internally held beliefs.

I believe in God. I also believe in the tenets and the structure of Islam. My heart and spirit are fully connected to the obedience of God. Prayer, diet, the notions of sin, even hypocrisy—all of this means most to me through the lens of being a Muslim. My Muslim-ness is at once anchoring and a way to make sense of the world. It's the source of my strength to deal with everything outside myself and fight for what I believe is right.

Yet I don't embrace a dogmatic view of Islam, the essence of which is interpretation. There are basic tenets, some shared by many religions, some specific to this one—but how you fulfill them is based on your own personal understanding. What does "Love thy neighbor" mean? And the commandment to read the Quran?

Fundamentally, true connection relies on using your own judgment. While posing certain challenges, this way also has its rewards. When you answer only to yourself and God, you don't need to worry about what others think.

"It's none of your business," I say to anyone who accuses me of not being Muslim enough. "I read. I understand. I am fully prepared to get judged for my sins and deeds on the day of judgment, because I will make my case. So you can back off."

Even in a realm as historically dogmatic as religion, I don't believe one size fits all. Faith is a pathway to life, but not just any life. Your life. My life. And none of us lives the same life. To wrestle with belief is to apply it within the context of your own circumstances.

That's why—although I had found showing up in the world as visibly Muslim by wearing a hijab to be internally rewarding—I didn't want Isra to wear one.

My daughter started asking when she was five years old.

"Maybe when you're seven, I'll buy you a hijab," I said.

I knew she thought wearing the hijab had something to do with being Somali, and I had a problem with that association. I didn't want her inheriting the cultural assumption many people have where a private choice turns into your presence in society. When the hijab is donned in this way, it's all about whether you are perceived as "religious" or not, a "real" Muslim or not. So you wear it at family events or holidays, and you take it off for school or to go to the movies.

I wanted Isra to develop a sense of what it means to exist as a Muslim outside of the markings. The weight of the hijab is not one that you can fully conceptualize as a kid.

Still, my daughter is nothing if not determined. On her seventh birthday, she woke me up in the morning and said, "Well, I'm seven today. You said I could have a hijab. Where is it?"

A promise, even a halfhearted one, is still a promise. I had to go find her a small hijab, which she wore for the day and then decided she didn't want to wear it the next. She asked Aabe to buy her a dozen hijabs with different colors when he made Hajj to Saudi Arabia in 2010, which he did. She wore them to holiday prayer, and so began her struggle with the commandment that so closely mirrored my own.

I would wear the hijab to prayers and other religious events. But when I was taunted for it in middle school, which led to more fighting, I felt the head covering was having the opposite of its intended purpose, so I took it off. Then I reversed course and wore it to school as an act of defiance. "No bully is going to make me take it off," I decided. Later on, the back-and-forth in my head became about feminism and what I wanted to reveal about myself. "I actually have beautiful hair and want to show it" as opposed to "In a sexist society, every woman needs a shield." All this wrestling expended a lot of energy. By the age of sixteen, I was exhausted and decided that no, I wasn't going to wear the hijab. That is, until I came full circle at the age of twenty-three, and I haven't taken it off since.

At fifteen, Isra arrived at the same solid place with the

hijab. "I don't want to wear this," she said. That's where she is, and I respect her decision—not just because I went through a similar process. I respect my daughter's personal relationship to God. I have no religious expectation of her or of anybody else for that matter. I'm a Muslim and live as such, but I'm also a humanist. Just as I believe in God, so also do I believe that we are all connected no matter our faith, belief in science, race, or country of origin. We all have an ability to enrich one another not in spite of our differences but because of them.

# 10

# EARLY MIDLIFE CRISIS

*2008–2009*
*Minneapolis, Minnesota*

As the presidential candidate Barack Obama was pushing the country to embrace hope and change, I was mired in hopelessness and completely at a loss as to how I could change.

In 2008, I was one of countless Americans caught in a deep existential crisis brought on in part by the huge structural flaws revealed by the financial crisis. The realization that my two-year degree, which I had fought mightily to earn, was as good as worthless in an economy where even elites were struggling left me disillusioned.

Meanwhile I had two children under five who needed so much. I felt like I was drowning.

Raised with the understanding that societal obligations are paramount, I believed in the pact that you cared for the weak, and if you found yourself in a time of weakness, you were in turn cared for. But that didn't seem to be happening, either in my family or in the wider world.

I struggled with who I was—both to myself and to those closest to me. I tried so hard to be the good daughter, the good mom, the good wife, the good friend. But I began to wonder whom I had been trying to satisfy all those years. What was I living for? It wasn't me. I had always felt like there were five hundred eyes watching me, and every single pair was looking to see if I met their expectations of who I was supposed to be and what I was supposed to do. When I got married, the number of eyes only grew, since now my family had gotten bigger.

I began to wonder about my own life. "If I was to die today," I asked myself, "could I honestly say I have lived a life that was worthy?" I had a whole list of things that I wanted to achieve that seemed further away than ever before, like getting a college degree. New desires started to bubble up, too. I had never really lived on my own terms, away from the confined space occupied by my family.

I questioned my most basic choices, including my relationship with Ahmed. Was marriage for me? What about my children? Did I have them because I wanted to or because it was expected of me? Similarly my religion, which had been a source of inspiration and strength, appeared in a strange light. What did I understand of Islam, really? I was unsure of everything that made me who I was. I couldn't say with any certainty that were I given the opportunity to make all my decisions again, I would make the same ones.

While I challenged the fundamentals of my existence, my surroundings grew intolerable—in part because I didn't think Aabe or Ahmed noticed that I was drowning. When they waved off my suffering with easy dismissals like "Oh, Ilhan, you're just tired," my disillusionment only grew.

**I BEGAN TO GET HEADACHES THAT WERE SO BAD I WENT FOR AN** MRI. Doctors couldn't find anything wrong with me. Still, I suffered from sleeplessness, pain, and a ringing noise in my ears. My insomnia grew so acute that Shankaron told me I had no place by the bedside of a sick relative in the hospital. "You can't be in the hospital unless you check yourself in," she said. "You look like you're the one who is about to die."

I had the strangest reaction to my two children. There were moments when I couldn't let them go, hugging them to the point where they would get scared and start crying. Other times I couldn't bring myself to interact with them, let alone embrace them. Dark, irrational thoughts ran through my mind; I could give them everything and they still wouldn't care if I died tomorrow.

It was as if nothing was real. I was emotional in ways that made me unrecognizable to myself. I wept when I got up to the cashier in the supermarket or to the server in the drive-through. Never one to open up much even to friends, I shared that I felt like I didn't have a purpose in life.

I was talking on the phone to a friend about how I was having trouble doing my job during this surprisingly challenging time when he interrupted. "What's going on?" he said, surprised by my oversharing. "I've never even been able to get you to tell me what you guys have for breakfast."

What was going on was that all the losses I hadn't spent any time mourning had come tumbling down upon me. Loss, for me, started with the most primal one: the death of my mother. In a show of bravado, I maintained that the funniest thing somebody could say to me when they heard my mother died while I was young was "Oh, how sad for you."

"Why would I need a mother," I answered defiantly. "My father is everything to me."

Of course, no one person can be everything, and the loss of my mother was just the start. The Somali civil war brought a string of losses. There were the mother figures I lost in her wake: the great-grandmother I never got to say goodbye to when we fled Somalia or my *habaryar* Fos, who died in the camps in Kenya from malaria. Then there were the other losses of war—the loss of my homeland, my language, and any sense of permanence. I lost not only my childhood but also the future that had been promised me.

At the time they occurred, I couldn't make sense of the enormity of my losses, so I ran away from them. But they were a huge part of my psyche, so in essence I had been running away from myself.

There were reminders, moments of yearning, like when I became a mother. Maybe I had talked to anyone who would listen about my first pregnancy because I was searching for someone who couldn't answer me. After I gave birth, when I didn't feel like I could talk to my father about private complications I was having with my body, I felt a flash of sorrow that I didn't have a mother.

Pity had never fit in with my identity before. I was always the strong person, the one who never cried. I was always the one who was sure of who she was, where she came from, and where she wanted to go. But that person didn't really exist anymore.

I couldn't communicate any of this to my family, in part because I felt like a fraud. "Ilhan is a rebel." "She marches to the beat of her own drum." "The baby of the family is the most courageous member." All these things I grew up hearing about myself couldn't be further from the truth. I wasn't bold. On the contrary, I was a coward.

I had suddenly discovered a great weakness I never knew I had in me. That weakness was allowing myself to mindlessly conform to family and social expectations without stopping to fully understand who I am and what my purpose is. I was the person I never imagined myself to be, the type of woman I railed against. When I considered my actions, I was somebody who was afraid of venturing out alone, being part of a relationship not accepted by my family, and parenting by myself.

If I were to know whether I truly possessed the strength others ascribed to me, whether I did indeed march to the beat of my own drum, I had to confront my deep-seated fears and see if I was actually courageous outside the confines of what "appropriate" looked like to others.

Life on my own terms.

My core.

My capacity.

**I WAS ANGRY WITH GOD AND WITH MYSELF, ANGRY WITH THE** universe and my family. Every single relative had a new face for me. Unable to figure out a way to reassemble a relationship with any of them, I was done. In this time of reckoning, I had to break with everything that was part of my confirmed reality. Because I was unsure of the love that my children, my spouse, and the rest of my family professed for me, my relationships with them were no longer binding.

By the summer of 2008, Ahmed and I were divorced. Because we had only been married Islamically (we applied for a marriage license, but busy with school, work, and family, we never got around to going to City Hall to make it official), all Ahmed had to do to end our marriage was declare it to be so. There was nothing formal. We didn't want to be together anymore, and that was that. It was such a subdued split, even our families didn't know Ahmed had moved out. I wasn't hiding our divorce; I just didn't care enough to share it. I was done sharing myself with people.

It was an emotionally chaotic time full of paradoxes. I was looking for a radical break with my past but followed my instinct to keep everything internal. There were no big outbursts or declarations. My breakdown was my business.

**THEN I HAD A BRITNEY SPEARS-STYLE MELTDOWN. I NOT ONLY** eloped with a man—whom I spent so little time with that I wouldn't even make him a footnote in my story if it weren't for the fact that this event turned into the main headline later on—but I shaved my head.

Yes, like the beleaguered pop star, who shaved her head in 2007, I took clippers to my own head. Too many headaches, too little sleep—I had to flee myself, my relationships, my hair. The difference between Britney and me is that I wore a hijab, so nobody knew what I had done—except my children, who were very surprised. "Mommy looks like me," Adnan declared. And he was right. I did look like a little boy. And like Spears's own elopement in 2004, my impulsive second marriage ended in divorce.

Despite the chaotic state of my mind, I had enough of a survival instinct to know what I needed to get better. I needed an injection of investment like Wall Street had been given in the bailout. I needed to invest in who I wanted to become.

I was compelled to search for a concrete goal to counteract the discomfort of feeling so alone and uncertain—and a new environment in which to achieve it. An undergraduate degree fit the bill. I had been on a long-term quest to earn my BA and decided this was the time to make it happen.

The opportunity came when one of my friends from high school, Daaq, told me that he was transferring to North Dakota State University, which was looking for students to enroll. The college was much more diverse than one would expect, since many Africans and students of other nationalities were lured to the campus by scholarships and a very low cost of living.

"What would I need if I wanted to go?" I asked.

"Just apply," he said.

And so I did. Without telling a single soul, I sat down at my computer and filled out the college application. Within twenty-four hours I received a response that I had been admitted and, not long after, talked to a school counselor over the phone to figure out the details of my enrollment. The next day, I called around and put a rent deposit on a place near campus.

My father and other family members were really worried as I loaded Isra, now six years old, and Adnan, now three, into the packed car bound for Fargo, but I felt great. To them, it seemed like I had lost touch with reality. It was true that I had kept them out of the loop on my decision to go to college in another state. I wasn't having a nervous breakdown, however. I was simply done with the opinions and advice of others. I was going to run.

# 11

# EDUCATION

*2009–2011*
*Fargo, North Dakota*

My alarm clock went off. It was four A.M.

Immediately I was out of bed, even though the cold of the North Dakota winter was unbearable. I didn't have the luxury of hitting snooze or even focusing on my discomfort. After wrapping myself up in warm clothing and a blanket, I sat down at the kitchen table to do my homework. The next few hours were precious time for me to concentrate on my assignments for the full course load I was taking plus all the extracurricular programs I'd signed up for as well.

Then the kids were up, and I busied myself readying them for their day. Isra was enrolled in elementary school and an after-school program at the Boys & Girls Club not far from where we were living. Adnan was also close by, in day care.

As my auntie Fay had said of children, "They bring their own blessing," and mine have always done just that. I can't imagine what my life would have been like if I'd had challenging kids. My children were simple, agreeable, and loving—so much so that back in Minnesota I often brought them along with me to work. A very chunky toddler who never stopped smiling, Adnan would run up to random strangers as if he had known them forever, and so people adored him. Even when he bumped his head or walked into a wall, he didn't cry. He just laughed. Years later, when I ran for a seat on the Minnesota state legislature, Adnan designated himself as the one to invite people to join our campaign. He was always a highlight at fundraisers and rallies, where he would remind people of why it was time for his mother to lead. Even when he was knocking on doors he won people over. "Please vote for my mother," he said while canvassing. "I know that she will take care of you like she takes care of us."

The morning light was just beginning to come up while I packed their lunches and snacks for the day ahead. Everything had to be ready before I bundled them up and got them out. I made almost everything they ate from scratch, spending Sundays cooking up the stews, grains, and porridges that I had planned as meals for the week.

I didn't have much money, so my motivation was partly economic. I also wanted to make sure my children, whom I had just uprooted and thrust into a whole new environment, had the nutritious food they were accustomed to.

My regimented food preparation was just one aspect of the environment I created in Fargo—where I was in control of everything as I saw fit. When I left Minneapolis, I was fleeing a family and culture that prescribed so many aspects of my life that I didn't know where they ended and I began. Away from the confines of their judgments, I wanted to explore the full range of my reactions—to new people, new books, new music, new ideas. In North Dakota I found a space where for the first time in my life I was able to push all of the expectations away from my brain and focus only on what interested me.

There was so much that sparked my curiosity. Joining any and every club that would have me, I was part of the black students association, the international students association, the college Democrats. Through my extracurricular activities, I became good friends with Jaylani Hussein, a natural organizer and community connector from Minnesota. We first met in an attempt to revive the Muslim student association at NDSU and ended up organizing the Muslim and Hijab Day to help educate people about Islam.

My coursework was equally politically heavy, which called into question my initial plan when I'd enrolled in college to get a BS in nutrition from the university's dietitian education program. That had been the logical choice based on the job I had had in Minneapolis for the previous four years teaching nutrition.

**MY WORK FOR SIMPLY GOOD EATING—A HEALTH AND NUTRITION** program for underserved communities through the University of Minnesota's extension department—began a few months after Adnan's birth. One of the members of our family ran a nonprofit engaging with the Somali community on a host of issues such as food insecurity. He asked if I would fill in for him as a volunteer, translating for the program's English-speaking teacher. He didn't trust the university to hire a Somali interpreter, he told me. A new mom at home, I jumped at the chance to get out of the house for several hours a week.

There I met Shelley Sherman, who instructed a group of Somalis on how to budget, shop for, and prepare healthy foods in America. Having lived abroad in Nicaragua and worked as a legal assistant to an immigration attorney in the United States, Shelley understood many of the issues facing immigrant communities new to this country. Many came from places where they were used to if not growing their own food then at least making homemade meals from fresh ingredients purchased daily in open-air markets. Accessibility to budget-friendly fast food and processed ingredients loaded with hidden salt, fat, and sugar means that everybody in the United States, no matter their background, has ample opportunity for an unhealthy diet.

I instantly hit it off with Shelley as she warned about the dangers of misusing the refrigerator and explained the high sodium content in canned vegetables. It turned out that she was about to become the program director, which left an opening for an instructor—a position that she eventually offered to me and I accepted.

A feminist and a caring human, Shelley put effort into building the self-esteem and potential of the young women around her, and she became an important mentor to me. She bucked the stereotype of toxic female bosses who are intimidated by and hostile to those who work for them. Thanks to her, I learned to develop very healthy working relationships with both women and men. She and other female bosses I had later on provided the model for how I, now a boss myself, interact with my staff. The big lesson was that it's possible to treat people as your equal even as you manage them.

On a weekly basis, Shelley solicited my assistance in adapting and fine-tuning the federally funded program's curriculum for all the different groups we taught. My classes were mainly in the English Language Learners schools, where I taught young mothers and expectant mothers who were attending high school. Occasionally

I filled in for the instructor who taught recovery groups, which was an eye-opener. As a visual illustration of calcium levels in various foods, we used different amounts of flour. But that, I learned, was a trigger for those in the room battling cocaine and opioid addictions. So we switched to cotton balls.

My favorite class was at the Lao Center, where the students were incredibly blunt. Culturally, the Lao people believe healthy eating means eating to full capacity. A rounded person was a pretty and happy person. Someone thin, like me, was sickly and unattractive. The men, especially the elderly ones, liked to tease me: "You are everything we don't want our women to be."

My experience with the nutrition program informed not only my concept of a productive work environment but also, perhaps more important, the connection of politics and policy to real people.

As part of the university's program, we were tasked with caring for and improving the wellness of the populations we taught. But figuring out how to purchase and prepare healthy food was only the start of their anxieties as they struggled with how to sustain themselves and their families in all aspects of life.

We were telling new immigrants not to keep their potatoes in the refrigerator or the starch would turn to sugar. Meanwhile, these same people were overwhelmed trying to learn enough words to fill out a job application or earn a driver's license.

There were the young kids, who were too busy putting their energy into baseline survival to care about the high-fructose corn syrup in soda. Some of them had parents who were incarcerated. Some of them were homeless. One day I was talking about the different nutrients and vitamins they should be consuming when a boy said, "Really? I don't even know if I'm going to eat tonight."

The federally funded program we were implementing was direct public policy. Those who represent us in government and create policy are oftentimes removed from the very people the policy is intended to impact. Therefore they don't understand the ways in which it's effective or ineffective.

The kinds of challenges that existed within these communities raised in me a new level of awareness of the larger systemic barriers that kept many of my students from being able to implement the information I was offering. The idea to provide nutritional education for better health outcomes doesn't work in a vacuum. Policy makers failed to consider policy in a holistic way.

I also shifted my thinking about the responsibility of lawmakers to their constituents. Yes, citizens have their part to play. When challenges persist in our communities, it's often because we don't know that there are public servants whose job it is to listen or how we can make them listen. Still, I came to believe wholeheartedly that no matter what, if you're worried about employment, housing, food, or healthcare, the politician who represents you should spend a lot of time worrying about those things, too.

**SO IT WAS THAT AT NSDU I FULLY PLANNED TO STUDY** nutrition—because it made sense for me to back up my professional experience with a matching college degree. I hadn't even really thought about it; instead, logic worked alone.

When I got to Fargo, however, in my process of questioning everything, I asked myself, *Why am I doing this?*

Every time I followed my instincts just for the sake of it and without the burden of justifying myself, I peeled back another layer that had been long covering the quiet core of who I am. While each step was emboldening, the greatest step toward true independence was deciding to pursue a degree in what I actually wanted to study.

In that moment of self-reflection I realized that my course load was heavily weighted toward politics, and that these classes were way more interesting to me than any of those I would have to take to become a dietitian. The most meaningful part of my job back in Minnesota had not been talking to people about nutrients but advocating for them. The latter was what would make me happy for the rest of my life.

At the end of my first semester I was talking to Baba on the phone, and he asked me what I had decided to study.

"Political science," I answered confidently.

Baba was incredulous. "Politics is a thing we do," he said, "not something you study."

I talked to my grandfather every Thursday. Like clockwork, he called first thing in the morning to ask if I was coming home for the weekend. I did spend weekends with him often, holed up in his house doing homework while the children played. Other than Baba, I didn't communicate with my family members, including my father and my siblings. Ahmed, a dedicated father, periodically came out to Fargo and stayed with us to see the kids. Baba, however, was the only one whose love for me I believed was unconditional and not predicated on my living up to a certain kind of image.

In my year of living for me, the rest of my family took a back seat. I had no interest in being emotionally there for them or keeping track of how they were doing. I didn't even seek out or spend time with the local Somali community in Fargo. I wasn't looking for connection with the familiar. I tuned out everything except the goal of getting my degree.

The separation from family was physically and emotionally painful, only masked by the grueling endurance game of waking up every dawn to do homework, making sure my children were taken care of, exploring my intellectual curiosity to its fullest, going to sleep, and repeating everything the next day.

I had always thought of myself as a liberated person

who believed in finding your center and following your internal wisdom. I didn't realize how my opinionated family confined and restricted me with their negative comments. Breaking away was the only way I could think of to find my core strength. I walked away from the experience of freedom and self-reliance understanding my full capacity and the source of my true happiness.

# 12

# RETURN

*2011–2012*

*Fargo, North Dakota, to Africa to Minneapolis, Minnesota*

"The only thing I have left is my work-study abroad," I told Aabe over the phone, "and then it'll be official."

I knew he was proud. A degree from an American university—it's what he'd always wanted for me. I was thirty years old, but I had earned my BA in political science and international studies from North Dakota State University. Well, almost. I still had to go overseas to finish.

"Maybe you can go to Somalia," he suggested.

Aabe always thought it was important to make at least one trip back. "In order to move forward," he had told all my siblings, "you have to make peace with your past and where you came from."

Then he suggested that we might go together. He was now retired from his job at the U.S. Post Office, where he had worked for over a decade, and he wanted to take a trip to our native country. Plus, he would help me with the expenses.

Before I left Minneapolis for Fargo, I would have seen his suggestion for the destination of my study abroad as meddling and controlling. Now, confident that I knew my own mind, I was able to say truthfully, "That's a great idea, Dad."

After designing a research project around the possibility that Puntland, a region in northeastern Somalia, might establish a one-man-one-vote policy like nearby Somaliland had done, in 2011 I embarked on a three-month trip with my father. We started out in Dubai, where I experienced my first Ramadan in a Muslim country. Then it was on to Djibouti for a couple of days, where I was fascinated to see its majority population of ethnic Somalis, whose home was another African country.

From there, we traveled to Puntland, where my father was born. The experience of going to a region I had never visited before on my first trip back to my native country since fleeing civil war was like waiting all year for a specific Christmas gift only to get a knock-off instead. Mogadishu was deemed too dangerous to visit, so I was in Somalia, but I couldn't visit my old home or school. I couldn't compare mental notes of what was to what is. I resigned myself to the idea of being home without really being home.

Meanwhile Aabe was having the opposite experience. It seemed that no matter what city street or rural field we were in, he found a memory. Dad would point to an empty space where a building once stood, a farm that had been developed, a well that had dried up, a tree that he had taken shade under as a boy. Happier than I had seen him since I was seven, Aabe smiled and giggled all the time. He introduced all those we met, from government ministers to taxicab drivers, to his "baby, a mother of two and college graduate, who has come back to our native country to have conversations about democracy."

In Puntland, where I stayed for a couple of months to research the viability of representational government, I learned that democracy can be achieved only if people are committed to the social contracts necessary to make it work. As I wrote in my findings, no amount of international influence or resources can make it happen without that. That's true in northern Somalia, and it's true everywhere in the world.

**MY EXPERIMENT IN LEAVING HOME FOR COLLEGE HAD SUC-**ceeded. I came away from my time at NDSU a much happier, freer, more centered person. In making peace with my past mistakes, I had become more forward-looking and optimistic.

Part of reconciling with my history meant reconnecting with the family I had run away from. I watched an episode of *The Oprah Winfrey Show* about forgiveness in which the talk show host said that most of us think about the concept incorrectly. The greatest burden is on the grudge holder. Often the person who is the object of anger isn't even aware they've done anything wrong. So in this way, forgiveness is about unloading the anger and sadness we carry. I was interested in forgiving family and those around me not just to forget, but so that I could let go.

During our travels, I had asked my father explicitly for his forgiveness. Remembering how angry and anxious he was with me during my teenage years, I felt like I had done something wrong to make him take on such a different persona than the liberal, democratic man I had known before. The protector and educator that I loved did not accord with the disciplinarian and judge I didn't much care for. But my dad's split personality came out only with me. "No man in my family has ever had a gray hair," he joked. "I'm going to be the first because of you."

If we were going to have a real relationship, I needed to reconcile these opposite sides of my father and say I was sorry for my role in creating this conflict in him. But that wasn't all. I wanted an apology from him for treating me like I had problems when I turned out okay. Actually, I had turned out more than okay.

My father apologized, too. I'm grateful for that conversation, which resolved our issues. I couldn't have imagined at the time how important my dad's unequivocal support would be for me when I later became a national—and international—public figure subject to relentless scrutiny and criticism for my behavior. To this day, and every day, people call, text, and email my father to complain about me. Then there are the online trolls and extremists who hate me. But there are also many members of my father's community who still hold him accountable for every photo, comment, or position they deem "inappropriate."

"They want me to put you in a box because you're a girl, who they think doesn't know any better," Aabe told me, "but I ask, how many people are calling the father of any male member of Congress to say, 'Your son needs a talking-to'?"

Even when his paternal worries kick in, which I know they do, he doesn't belittle me with protectionist admonishments. "That's my daughter, and I'm proud of her," he says. "She is a full being. She gets to have autonomy over her decisions and how she wants to live."

**IF THERE WAS ANY RESIDUAL HURT OR ANGER AT MY FAMILY** and the choices they had made for me, my three-month stay in Africa brought a new level of appreciation for the challenges they had faced. After leaving Puntland, I traveled to Nairobi to interview some members of parliament. The Horn of Africa was experiencing the worst drought to hit the region in sixty years. Millions of Ethiopians, Kenyans, and Somalis were starving. So my plans changed.

My detour started in Nairobi, where I connected with young Somalis I knew from Minnesota, who were planning to visit Mogadishu. At the time, I was under the assumption that my native city was too unsafe to visit because of the rise of Al-Shabab, a violent insurgent group that had claimed the capital. My friends told me that while parts of Mogadishu were under the control of this organization, other areas were controlled by the government and safe enough for the Somali diaspora to visit.

Excited by the prospect of going home, I inquired about places to stay and how much money it would cost for me to get there. They were leaving the next morning at five A.M. and would help me find a safe place to stay. Just like that, I was on my way to Mogadishu. (I neglected to ask or tell my father until later that night when he wondered why I was packing—I didn't want him to worry or try to stop me.)

When I arrived in Mogadishu, it was not the city in which I had lived. No monument was fully intact. Familiar roads were blockaded. Both my great-grandmother's house and my childhood home were inaccessible. Grief-stricken, I went back to the hotel and fell asleep—for sixteen hours! When I woke up, it felt as if a week had passed. I went down to the hotel restaurant, where I engaged the red-bearded older man who had worked there for decades. I told him that I couldn't seem to wake up since arriving in the city a day and a half earlier.

"You slept like a child who had been away from home," he said. "You always sleep with ease in the place where your umbilical cord is buried."

The man's statement worked immediately to restore me. I felt understood and at home.

**I STAYED UP LATE TALKING WITH THE ELDERLY GENTLEMAN** about what was and what is. When I asked what my chances of traveling to see my childhood home or any sites familiar to me were, he said it wasn't possible. "Be happy that you are able to set foot in *xamar cadey*," he said, calling Mogadishu the name used affectionately by locals.

A few hours after going to bed, I was awakened by singing and dancing. A jubilee filled the streets. I stumbled half-awake down the stairs, confused at the sound of music and shouts until I learned that, at the crack of dawn, Al-Shabab had retreated from the city. It was like a fairy tale.

I quickly went in search of the elderly gentleman to

ask again about seeing my childhood home. For fifty dollars, he said he could arrange for men to drive and guard me—an offer that I agreed to immediately.

Not much was left of our home. I stood next to one of the only standing pillars in what used to be our living room while scenes from my childhood played before my eyes like an old family video. There was the outline of the great room I would burst into as a little girl after I had outrun the classmates I had hit back with sticks from the bushes outside. I smelled the cakes baking near the corner of what had been the kitchen. I recognized the place where Farah parked the Corolla.

I wept until the men paid to be my escorts reminded me that there was always the possibility that Al-Shabab wasn't wholly gone. It was time to go back to the hotel. I didn't leave, though, until I took a few pictures to bring back to Baba, who, until he saw the pictures, was convinced our home still existed as it had when we left it.

The next day I said goodbye to Mogadishu. I had to return to Nairobi and my worried father. At the airport, I saw a familiar face. It was Abdisalam Aato, an established Somali filmmaker. He didn't know me, but that didn't stop me from sitting next to him, introducing myself, and starting a conversation. He was either alarmed or not interested in talking to me, because he excused himself to go to the restroom and disappeared. Undeterred, I sat next to him on the plane, where there were no seat assignments. It wasn't until we landed at Wajir Airport in

With my family in Mogadishu.

Age 10 in Kenya.

With my dad in the 1990s.

With my cousins in Canada in the 1990s.

Campaigning against voter restriction measures in 2012.

My 2014 DFL Women's Hall of Fame
Rising Star Award ceremony.

My last day in Andrew Johnson's office.

Campaigning at the University of Minnesota in 2016.

A meet-and-greet at a University of Minnesota coffee shop.

During the 2016 DFL bus tour across Minnesota.

Candidate filing day, 2016.

Election day, 2016.

Team Ilhan at the 2016 caucus night celebration.

With State Senator Scott Dibble at the vigil for the Pulse mass shooting victims in 2016.

With Erin Maye Quade at the Pulse vigil.

At the 2017 Women's March.

January 3, 2017, with my campaign team on the day of my swearing-in to the Minnesota house of representatives.

Making a cameo in the Maroon 5
"Girls Like You" music video.

Speaking at a pro-immigration rally in 2019.

On the day of my swearing-in to Congress.

Ilwad got sworn in with me.
*Photo by Glen Stubbe/*Star Tribune *via Getty Images*

Signing my oath of office.

My first time presiding over the United States House of Representatives.

With Speaker Nancy Pelosi in Ghana at the Door of No Return during the 2019 congressional trip to Africa.

Kenya, where security checks all the planes from Somalia before they can go on to Nairobi, that Aato finally opened up.

As we sat waiting for our luggage to be screened by bomb-sniffing dogs, he told me he was wary of travelers in a land where you couldn't always be sure why someone was trying to be friendly. Then he explained that he had gone to Mogadishu as part of a youth movement of Somalis from around the world. They had collected money from the diaspora to purchase supplies like water, glucose, and sandals in Nairobi that they then planned to deliver directly to Somalis facing starvation on Kenya's northern border. I asked if I could go with him. "If you're available to go to a meeting after we land," he said.

The week that I spent helping people fleeing famine across the border from Somalia into Kenya was life-altering. As a refugee who has escaped tragedy, you never stop thinking about all those who didn't make it out, how much they suffered and how many died. To go back and be of service to people in the same position my family and I were once in helped me find some peace. I didn't just leave, get a better life, and never look back. I was here and trying to help.

There were so many people, families, and stories from that experience that have stayed with me, but one woman gave me particular insight into and empathy for my family when we fled Somalia. Jija had lost five children and her husband on the journey to the makeshift camp on the border where I met her. But she had three smaller children who survived because they drank cooled urine. The older ones, who wouldn't, died.

"How did you know they were dead?" I asked.

"Once they couldn't walk anymore I had to leave them," she said.

I was instantly gripped by the horrifying thought of the hyenas along the road. How does a mother walk away from her child, unsure if they are dead but knowing that even if they aren't, wild animals will eat them alive?

Almost as quickly my shock was replaced by a deep gratitude for what my father, grandfather, aunt, and other relatives had to endure during our journey to the refugee camp. For a long time I had been angry that we hadn't all traveled together. The adults in the family made lots of choices during that time, many of which never made sense to me. In that moment, though, I realized that they were singularly focused on survival, no matter what. Listening to this mother and being a mother myself, I could finally understand: you just have to keep going.

As I traveled back to the United States, I kept thinking about the greatest privilege I had been afforded—I had been allowed to live. I was overwhelmed with a need to show my appreciation not only to the city that adopted me but to the country and world that came together in order for me to survive. I felt obligated to return and speak about my refugee experience for the first time and to advocate for empathetic policies that take into account real human suffering.

But first I needed to find my way back home.

**WHEN I HAD LEFT FOR NORTH DAKOTA, BABA'S ONE REQUEST** was that I return home to Minneapolis. Although he lived two more years, he was convinced his time on this earth was coming to a close—and he wanted to make sure as many of us as possible were around him when the day came. "I had the misfortune of burying many of my children and their children," he said. "I don't want to have the misfortune of not having any of them bury me."

Before I returned to Minnesota, having earned a BA in international studies and a BS in political science, I wanted to complete the exercise I had done with my father—to forgive and ask for forgiveness—with all my family members. So that is what I did. To some I wrote letters. To others I wrote emails. I even contacted a few through Facebook. They allowed me to come home to Minneapolis without a single sadness. I had a complete rebirth.

As those who feel better about themselves generally are, I was now a better person to be around. My siblings were content with their sister who wasn't chaotically screaming at them about every single one of their opinions.

Ahmed and I found common ground in the same famine relief work that I had first encountered in Kenya. Somalis in Minnesota were a huge part of the famine relief support, and Ahmed was one of its main organizers. Working collaboratively on this important initiative, we connected in ways we hadn't before. As we engaged in saving Somalis, we actively worked to save our family.

Reflecting on the hard choices people had to make for survival back in Africa, I had to reconcile the reckless decisions I had made for survival of a different kind. The breakdown of my marriage, like that of my relationships with other family members, was a direct result of my unresolved conflict over the fact that while people and places I loved had been destroyed, I had survived. I had to learn to forgive myself and fully accept the woman I had become, before Ahmed and anyone else I loved could do the same.

Ahmed and I repaired our marriage, but our relationship had changed. He had to come to terms with my need to exist fully as myself, including as a wife and a mother. The shift was at once subtle and enormous. It came in little moments that would have quickly devolved into fights or grudges before I left. If Ahmed got up in the morning with the children and made them breakfast while I slept in, he remarked on it when I woke up. In the past, I would have blown up. Instead, I recognized that I didn't have to feel bad about sleeping an extra hour. It was very freeing.

I also questioned the gendered aspects of parenting. "Isra's hair needs to be fixed," Ahmed said, looking at me.

"Okay," I said, returning his gaze, my meaning clear: *So, do her hair.*

My response wasn't hostile, it was just a fact. I wasn't going to fix my daughter's hair simply because I am the mother and a woman. Ahmed was just as capable as I was at doing hair. In fact, I was terrible at it, perhaps as a reaction to growing up without a mother. Either way, Ahmed did Isra's hair without a fight, because he is an evolved man, one who realized that breaking the stereotypes of what was expected of each of us simply because of the sex we were assigned at birth was freeing for him, too.

Ilwad—our third child and the gift we needed even though we didn't know we needed her—was born in the summer of 2012. As she got older, her siblings would tease her by saying things like "You were an accident."

Ilwad wasn't bothered by this. "Yes, God sent me to make you guys happy," she would retort.

She was like the proverbial guest in the holy texts, someone God brings into your home to inspire and give you life. She made all of us happy, because there isn't a space Ilwad occupies where people are not happy. Even as she grows, there's something about her that remains transcendent.

Her name means "to see beauty" or "the beautiful" in Somali. There is no direct translation in English. Isra picked out the name, but I was uncertain about it while

I was pregnant. "What if she's not beautiful?" I asked my father. But even Adnan, who was really upset right after she was born since he had wanted a brother, had to admit when he first saw her, "Fine, she's the most beautiful thing I've ever seen."

# 13

# POLITICS

*2012–2013*
*Minneapolis, Minnesota*

I was nauseous and hungry, tired and cold. But time was running out, so even though I wanted to go home, put my feet up, and eat Thanksgiving leftovers, I returned to the elevator bank and pressed the button to go up.

In the two months that I had been doing get-out-the-vote work for Mohamud Noor's bid for Minnesota state senator from the 59th district in a special election, I had been to the Riverside Plaza apartment complex more times than I could count. Some of the mostly Somali residents were surprised when a pregnant hijabi knocked on their door and wouldn't leave until she had convinced them to vote for Noor in the Democratic primary on December 6, 2011. Or at least vote.

I hadn't been home from Africa long and was already pregnant with Ilwad when I threw myself into the campaign that ignited me and many others seeking progressive representation to bring about economic prosperity and opportunity for all of Minneapolis's residents.

Noor—a Somali immigrant, a former employee of the state's Department of Human Services, and a political newcomer—was running against Kari Dziedzic, a policy aide to the county commissioner and daughter of legendary longtime Minneapolis city council member Walt Dziedzic. Carrying on a political family dynasty, Dziedzic was pushed by the Democratic establishment while Noor, running to become the first Somali in the country elected to a higher office, was of the people. At least that's what I thought and what I was willing to work for.

Noor's run for a seat on the state legislature was the first campaign I volunteered for after returning from North Dakota, but I had been involved in many forms of community organizing way before that.

My interest in the workings of our democracy began a few years after I arrived in the United States, when Baba asked if I would accompany him to a caucus. He had heard about the gathering in an English class he was taking at a local resource center. Understanding it to be some manner of participating in democracy, he was interested in attending and wanted me to make sure he was able to follow along with what was being said and done.

Neither of us had ever been to anything like it. Our first introduction to the event were the people hawking their candidates outside, trying to get passersby to come into the venue like carnival barkers for public policy.

I was able to translate the words for Baba, but I couldn't make sense of the process. At first people were just sitting around, and then, without any kind of proper introduction, someone eventually started talking, only to be cut off by someone else yelling, "That's not how you do it!" A little later, an agenda was handed out about issues to be discussed. Then people began to argue about the agenda. I don't remember the candidates, but the room was filled with a lot of characters. It was a good place for people watching.

Even though the whole thing was messy—*really* messy—it was clear that everyone was eager to be there. As much as they were screaming at one another, there was a passion for the process that even I, then a fourteen-year-old, could appreciate. No one voice was more important than any other at the caucus—anyone who showed up could participate to their fullest.

That seminal event stayed with me not only because Baba and I continued to attend caucuses together. Seeing democracy in action at the caucus was what convinced me of the power of the vote. It's the cornerstone of our system. Everything else builds from that one act.

My studies in political science and my efforts organizing volunteers for campaigns and causes only solidified this viewpoint. It was always my belief that the first, most important step was to get people to vote—not because of a specific issue they felt strongly about but simply because it's our right. My mission in my canvassing work was to help people understand that it should be an honor to cast your ballot.

I disagree with those who argue that educating voters on the issues or candidates is crucial to the process. Worrying who is going to vote and why is an afterthought, a privilege. In the United States, one of the greatest democracies in the world, we have one of the lowest voter turnouts, which is a disaster and a disgrace. I can express the importance of voting succinctly. It is not something that needs a long explanation. We vote because it's a right enshrined in our Constitution, and one that we must exercise.

I brought this message to hundreds of residents during Noor's two-month campaign for the state senate. Ultimately, he didn't win. But we put up a very good fight.

In a five-person race, Dziedzic—who had raised the most money and had the most support from elected officials and unions—beat Noor by only six points. As the *Star Tribune* reported, "His second-place finish was marked by a dramatic turnout from the city's Somali community." Ninety-five percent of voters in the heavily Somali precinct surrounding the Riverside Plaza apartment complex cast their ballots. "No other precinct turned out more voters," the newspaper added.

As it turned out, Dziedzic became one of Minnesota's greatest senators. There's a narrative around a politician and then there is reality. As someone who worked hard for her opponent, lived in her district, and served alongside her, I'm uniquely qualified to say I was wrong when I thought she wouldn't be a true representative of the people. Sometimes the terms *liberal* and *progressive* are not rooted in work and efficacy but in the people affirming them. When Dziedzic won, we didn't expect her to engage the community. But her ability to do just that and to advocate for policy change has been tremendous. The fact that her campaign against Noor had been so rigorous and close might have also helped shape what kind of senator she ended up being. Where once she might have been persuaded to adopt the status quo, she now had to serve a district that wasn't going to settle for that.

**AFTER I RETURNED TO MINNEAPOLIS WITH MY COLLEGE DE-**gree, I did what I had done when I had arrived at NDSU—I just signed up for everything and anything in the progressive movement that piqued my interest. I joined the board of many progressive nonprofits and was involved with a political action committee that did work in the new American community to get politicians to care about our issues. I was also elected the vice chair of an organized body of volunteers for my senate district, which Dziedzic had won.

I was everywhere, although I was never the person out in front. My name wasn't on the agendas or the billboards bearing policy initiatives; I didn't offer comments for the paper on community meetings. I was a behind-the-scenes busybody. Unafraid of confrontation, I was sent to meetings to upset people and hold them accountable. But I also planned rallies and chaired conventions. I joined Minnesota Young Democrats and made resolution recommendations at the state central committee. I am the kind of person who truly enjoys spending ten hours on a Saturday talking through the party platform.

If there was a need, I showed up—with or without permission. People would often ask me, "In what capacity are you here?"

"Why does that matter?" I would answer. "I'm here."

I always thought it odd when people created obstacles for themselves that got in the way of addressing an issue or being in a certain space. My guiding principle was "Do I care enough?" If so, then I didn't care what anyone else thought about my presence in a room or conversation.

In 2012, the Republicans, who had a majority in the state legislature at the time, drafted two ballot measures: one was to require photo ID for voters, and the other one was to define marriage as between one man and one woman in the Minnesota constitution. I immediately jumped in to get the vote out *against* the two referendums.

"Be Nice, Vote No Twice," the grassroots campaign against the two constitutional amendments, was exciting for me because I got to speak about my love for democracy and voting.

Although the voter ID measure predominantly affected people of color, as all laws aimed at voter suppression do, it wasn't the not-so-subtle racism of the proposed amendment that had me fired up. My primary objection was bigger than that. It went back to the idea that voting is the right of any American. You don't need to pass a test, understand what you're voting for, or have a license. If you are a citizen, you get to vote. It's as simple as that.

The state Democratic Party, running the advocacy against these ballot measures, determined the overall messaging of the campaign. The immediate issue we faced as the grassroots volunteers charged with implementing the centralized message was that our community members didn't necessarily respond to the conventional narrative the campaign had put forth.

The argument the campaign put forward against the photo ID measure revolved around access obstacles to obtaining driver's licenses. Perhaps in other communities or other states, it is hard for people to obtain a driver's license. Not in Minneapolis, where whether you were white, Latino, African American, East African, or a student at the university, the barrier to getting an ID was pretty low. The Department of Motor Vehicles is centrally located and accessible by public transport. And the cost is low. So the campaign's argument just wasn't the right approach for our community.

The statewide mandated narrative against the marriage equality ban was modeled after President Barack Obama's nationwide campaign. It encouraged people to connect and share their stories about LGBTQ friends and loved ones. In many of our marginalized communities, the conversation around "I'm your friend. I'm your son. And because I am, you should support my right to marry who I want" wouldn't resonate.

I felt it was my responsibility to figure out the right narrative and the right tools for when we reached out to our districts. Because I knew what would work with me, I was confident I knew it would work with the people around me. Tailoring a centralized directive to each community is the task of those on the ground in a huge grassroots effort like this was.

The winning narrative—that these measures threatened freedom and civil liberties—wound up speaking not only to communities of color in Minneapolis but also to white rural Minnesotans. For greater Minnesota and more conservative rural regions, the argument was if you vote yes, you're infringing upon people's rights by changing the constitution, which is intended to protect freedoms. Voting no, however, means nothing changes. For urban and immigrant populations, the same idea based on the importance of preserving freedom was slightly modified: if today they come for one community, such as same-sex couples, who's to say they won't come for another community, like yours, in the future?

This clever and clear messaging that worked everywhere was driven by values-based organizing. We didn't try to change how people felt. Instead, by explaining that the amendments threatened freedom and civil liberties, we activated citizens to vote based on their core values.

The result was a defeat of both ballot measures. This was no small feat, as no amendment defining marriage as between one man and one woman had ever been defeated on the ballot in the United States before. Minnesota became the first in a string of states that later defeated similar proposals, paving the way to the legalization of same-sex marriage in all fifty states three years later. The groundswell of organizing made 2012 a ridiculously good year for progressive Minnesota. Democrats, who already occupied the governor's office, took back the House and the Senate as well that year.

**BY 2013, I HAD GAINED A REPUTATION IN THE SMALL WORLD OF** progressive Minnesota politics for being a hard worker and effective in expanding the platform of our party by making it more accessible. Perhaps that's why Chris Meyer, part of this small community of young activist types, floated my name for campaign manager for Andrew Johnson's bid that year for an open city council seat in south Minneapolis.

Meyer was acting as interim campaign manager for Johnson, who had the worst odds out of all the candidates vying for the seat representing the 12th ward. When Johnson approached me, I wanted to say no. I was just coming off the "Vote No Twice" campaign, which had taken up a lot of my time and energy. With three small children at home and Baba very ill, I didn't want to enter into another consuming political battle.

Johnson was a twenty-nine-year-old former Target systems engineer, and I knew from our time together at my district's neighborhood association that he was the kind of person who put in the work without requiring any attention for it. As the youngest member on the city council, he would be a good addition, since he cared about the issues and not just the politics.

For a long time, the Democratic-Farmer-Labor Party (DFL), Minnesota's center-left political party, didn't have young people in its ranks who were ideologically driven to move its platform to the left. The leadership and those it attracted consisted mostly of older white cisgender and straight citizens when the majority of the state in actuality had shifted to young, disabled, brown, gay, and whatever other categories made us a truly diverse state.

Johnson was part of a larger resurgence in representative democracy in the Twin Cities.

The excitement around the election of a coalition of insurgent candidates (including Johnson) who were challenging long-standing incumbents in wards across the city was enormous. Suddenly the issues we had been raising—such as disparities in infrastructure, education, and other municipal investments in neighborhoods of color—but never really thought would be addressed on a massive scale, were on the table—if we could get our candidates elected. A cohort of young, ideologically extreme-left organizers signed up not only to work for particular politicians running for office but also to revitalize our party and its constituent caucuses—the Somali and the Stonewall caucuses, just to name a few. As it turned out, seventy-year-olds weren't the only ones who liked to sit at a Perkins coffee shop forever, drinking coffee and hashing out policy reforms.

So I signed on to run Johnson's campaign, joining a group of young repeat serial DFLers like myself. Chris, David Gilbert-Pederson, Wil Sampson-Bernstrom, and I made up the ragtag team in a race that turned out to be even tougher than I imagined.

**AT FIRST, JOHNSON'S BIGGEST HURDLE WAS BEATING A SIXTEEN-**year city council incumbent, Sandra Colvin Roy. So when Roy, in the face of Johnson's growing popularity, withdrew her bid for reelection in June, those in the progressive movement took an early victory lap.

Those of us on his campaign, however, quickly understood that Roy, who backed one of his opponents in the race, was still actively trying to undermine him every single day.

Busy devoting themselves to the slate of other progressive candidates trying to unseat incumbents, the progressive movement leaders responded to our pleas for help with "You guys are fine."

"There's the whole organized machine to make sure Andrew doesn't get elected," I argued. "How can you not see it?"

That meant Andrew; Wil; my deputy campaign manager, Chris; and I were working twelve-hour days, seven days a week, strategizing, writing and producing our own campaign literature, writing speeches, prepping for debates, and knocking on doors. Campaign managers don't typically go out knocking on doors, but I had to do a lot of it. All four of us did, because there was no one else but us. Chris, who moved into Andrew's house, was knocking on doors more than any human ever should have. I felt so bad for him, but he was a natural.

Our challenge wasn't just manpower. It was also a very dirty campaign full of constant thuggery. Signs we put up on lawns went missing, as did literature that store owners had allowed us to locate in their businesses. The lowest moment, however, was when someone dredged up two blogs Johnson had written when he was a teenager that threatened to derail his entire candidacy.

It was October, and in the run-up to the election, Johnson was predicted to win. That's when we were blindsided by these blogs he had written when he was an adolescent about such subjects as his drug and pornography use, cutting himself, and making rockets with explosive chemicals. It was bad.

Johnson, who later called it "the hardest moment of my life," felt violated that a part of him that he no longer saw as himself was being used to ruin everything he and the rest of us had worked so hard to achieve.

We needed him to function. I worried that reliving what must have been a dark time for him as a teenager would be too traumatic for him to keep it together. So I stayed away from questions of how and why he wrote the blogs and focused on trying to get him to eat and move on.

"We have your back," I said. "We're going to knock on doors and do everything we can. These people are interested in making sure you don't win, so you can let them by staying in bed or you can prove them wrong."

I walked him through what needed to happen, such as putting out a statement and making calls to the elected officials who had endorsed him, including the mayor of Minneapolis. "The rest is how people react," I said. "We'll see."

The question was whether he should talk to the *Star Tribune*, which had published articles that cited the original blogs. He could go on the record and try to explain himself, or they could spin the story whatever way they wanted. Neither choice was great.

He was hurting and angry at these people who had put his deepest darkest thoughts out there for all the world to read. The editors were on the opposing side, but we had to think strategically. It was so close to the election that ultimately we decided to go on the record. There wasn't enough time for the story to fade away.

When we sat down in the newspaper's conference room filled with editors to address the controversy, I could hear Johnson's stomach gurgling with anxiety. I wasn't moving from his side. We had gone over his talking points many times. And if they asked anything inappropriate or he felt like he was going to cry, I had told him: just look at me.

I had my own emotions I needed to control. Angry at how crappy they were to think this was a story, I wanted to scream and hit them with something. They sat there stone-faced, not betraying a single ounce of empathy. Meanwhile, Andrew looked like he had been run over by a truck. He was visibly shaking and drinking way too much water.

That meeting altered my relationship with the media. It wasn't just their willingness to exploit vulnerability in order to sell papers that disgusted me. It was also the way they were talking to Andrew, as if he weren't a human being.

He got through it, and the story came out. It wasn't

a love letter, but it seemed that people understood he had been a kid when he wrote the blogs. Still, the timing of the article—less than three weeks before the election—was terrible.

On election night, we assigned volunteers from the campaign to each precinct. Andrew, Wil, Chris, and I had all gone to precincts that were not predicted to be winnable since those are the ones where you can get a lot of good intel on how your candidate is doing. At my precinct, they read out the numbers, and Andrew was clearly ahead. I immediately called Andrew, who told me that he had won his precinct, too.

I needed gas before I could drive over to the campaign election-night event. On my way to the gas station, I did the calculations and realized that even if we didn't win Wil's and Chris's precincts, we had won the election. I started to cry. When I got out of the car to pump gas, Andrew called.

"They just called it," he said from the brewery where the celebration had already started. "We won."

I was so excited and happy I started to scream. We had given everything in time, spirit, and political capital to this brutal campaign. The catharsis from our victory was overwhelming.

Without thinking I dialed Baba's number to tell him that we'd won. But instead of his low voice, I got a digital one saying, "The number you have called is no longer in service."

I started crying again, but this time out of sadness. My grandfather had passed away a couple of days earlier. He was always my first call in the best, worst, and stupidest moments in my life. I had completely forgotten he was gone.

**BABA'S DEATH WAS THE FIRST TIME I HAD EXPERIENCED THIS** depth of loss since my *habaryar* died in the Utange camp. I longed to hear his voice again, to listen to him speak once more. Baba's language was always indirect, even though he unfailingly made himself perfectly understood. He was a very forthright man, but not in language. In Somalia, known as a nation of poets, it is a sign of intelligence to communicate through metaphor. Baba was a master at making his strong opinions known through the gentlest proverbs. Calling him was a reflex that would take me more than two years to lose.

I walked into the gas station to pay for the fuel. There was a guy standing there, and I hugged him while crying and laughing at the same time.

"Are you okay?" he asked, obviously concerned.

"I just tried calling my grandfather, but he's dead," I explained. "I wanted to tell him that all my hard work paid off."

# 14

# CITY HALL

*2014–2015*
*Minneapolis, Minnesota*

Although by 2014 I'd spent a lot of time in politics, City Hall was a completely different beast.

As senior policy aide to Councilmember Johnson, my first full-time position in government, I had to open his office, hire staff, and get him prepared for his swearing-in. The bigger challenge for me, however, was the loss of autonomy involved in working for an elected official.

Before I went to work for Andrew, I had never been a paid staff member of any organization. I reported to no one, and no association kept me from expressing my opinions and ideas. Even if someone wanted me to back off or check myself, there was literally no one for them to enlist in the cause. I existed as my own entity, free to ally myself with whomever I chose.

Now my opinions were Andrew's opinions. My political advocacy was considered Andrew's. So we had long conversations in order to come to consensus. We had the same overall goals. For example, we were always pushing to get rid of outdated ordinances. Andrew and I sifted through thousands of them, looking for ones that no longer served their purpose. People began making fun of us, because every month we introduced the cancellation of thirty or so ordinances. It kind of became our brand.

However, there was one regulation we disagreed about getting off the books: the prohibition against spitting and lurking. I considered this rule worse than an outdated law—I thought it targeted young black and brown men. Citations for spitting and lurking weren't the problem. The ordinance was a gateway to police harassment. The excuse to stop men of color was spitting or lurking. But if these men questioned why they were being stopped, officers escalated the offense to resisting arrest or carrying drugs. Normally you can't just pull over and search someone who is standing on the street. If lurking is against the law, though, police suddenly have authority to do just that. I argued that standing and lurking have a distinction without merit.

Andrew wasn't convinced. "There's no data to support the fact that black and brown men are being harassed because of this ordinance," he said. "You've looked at it backward and forward, and you cannot find a connection, so I don't understand why this is a problem."

He was right. I had tried everything to find "proof." But inherent in the unjust system was that the proof didn't exist. All I could do was ask him to trust me.

"That's our lived experience," I said. "It might not be for you. The data might not exist. But I'm telling you."

Organizers I knew who worked in the area of justice and racial equality—friends with whom I had organized rallies—questioned my commitment to progressive values. "Why are you working for this guy?" one of them asked me.

Navigating this issue was part of a larger test of how activists can fully inhabit a role where they are the ones being pressured by different groups. How do you maintain your ideals and outspokenness while remaining respectful to your colleagues?

What my criminal justice advocacy friends didn't understand was that Andrew and I had different processes for coming up with an answer. He appreciated my process and I understood his. It was on both of us to convince the other one, without invalidating anybody's viewpoint. Tricky stuff.

Because of my personal history of dealing with oppressive systems, almost as soon as someone brings up an issue of inequality, I'm on board. I can get there pretty fast. But the passionate argument was not one that worked for Andrew, who was a numbers person.

I didn't mind doing the extra work to make my case, whether through data or perseverance. Through our dialogue, I learned how useful it is to have both sides in the decision-making process. To counteract getting swept up in emotions, logic and concrete proof are crucial in judging an issue on its merits. On the other hand, inequality can be so deeply entrenched that sometimes finding proof entails questioning the very premise of the system. That's why it's always vital to return to the humans and be able to hear their stories.

In the end, I convinced Andrew to cancel the spitting and lurking ordinance. He didn't regret his decision, and not only because I put the work in to make my case, even if it wound up being mostly anecdotal. Andrew also trusted me, because he knew whatever counsel I offered him, I would take responsibility for it and be right by his side while he implemented it.

One time we heard rumblings that the union was accusing the city's Department of Civil Rights of enabling a toxic work environment. Andrew asked me what I thought he should do.

It was a sticky situation. The department was run by a woman of color. But this woman, whose job it was to litigate violations of people's civil rights on behalf of the city, was being accused of abuse by her staff of all shades and backgrounds. Enter a white man—Andrew—whom everyone was coming to with their complaints. What was he supposed to do? If he came down against their boss, he opened himself up to being called racist: "Of course you don't think the black boss knows what she's doing." Or sexist: "Whenever a woman is a tough boss trying to make sure that people are doing their jobs, she's abusive."

When Andrew broached the issue with me, I think he fully expected me to talk him out of dealing with it on the basis that he was walking into a political land mine. Instead, I said, "Let's suit up and head on down there."

I didn't think that we could afford to ignore the problem just because he might face charges having to do with race and gender. You can't decide you're not going to engage in an issue because it's too controversial. In fact, often the most contentious and difficult problems need the most attention and discussion.

"Why don't we surprise them to see how the place works?" I said.

He was skeptical.

"I'm with you. Right?" I said. "We're a team. People know you're actually interested in facts—and that I don't take bullshit."

Most important, these were city employees, so we had a responsibility, even if only indirectly, to make sure that their workplace was productive and safe, and free of discrimination. I suspect Andrew felt the same way and just needed me to say it out loud, because we put on our jackets and headed straight down to surprise a department of municipal employees.

People looked very confused when we walked through

the rows of cubicles. Some engaged us right away, and we told them the purpose of our visit: to ask employees about their experience working in the department. Others quickly looked away or flashed an expression that could be read as "Please don't talk to me, because if you do, my life is going to be a living hell."

We didn't gather enough useful information that day or in the months that followed for us to fight for them properly. Work-related disputes are extremely challenging to adjudicate. Ultimately, Andrew was confident enough to vote against the reappointment of the department head—not before getting chewed out by the council president for visiting the office uninvited, however.

**IT'S MY BELIEF THAT IN PUBLIC SERVICE, IF YOU AREN'T MAKING** someone uncomfortable, you aren't doing your job. And from the time I was old enough to go to school, I never shied away from conflict—even if it was physical. I just never imagined that I would experience a conflict in politics that would turn physical. But that's exactly what happened on February 4, 2014, at a caucus in the race for a seat on the state legislature.

The race for Minnesota House District 60B between incumbent Phyllis Kahn and Mohamud Noor had already seen an unprecedented amount of ugliness. It was the toughest fight Kahn—elected to the state house in

1973 as part of a grassroots effort to bring women into what had been a boys' club—had faced in her forty-two years as a state rep. Over the previous two decades, the makeup of Kahn's district had changed radically to include an influx of Somalis fleeing civil war, like my family. Naturally, some Somalis began to wonder when a member of their own community might represent them. Noor, now a Minneapolis school board member, was the first real challenger to go up against Kahn.

Like all communities, the Somali community is not a monolith. There was a division between progressive upstarts looking for change and the establishment, which had formed an alliance with Kahn, a powerful political operator in the House. Minneapolis city council member Abdi Warsame fell firmly in the latter camp. A year earlier, Warsame had become one of the first Somali Americans in the country to win a municipal election when he was voted to represent Ward 6. He had no interest in seeing the political system upended—even for another Somali.

The day before the caucus, Andrew received a call in his office. It was Warsame, and he didn't mince words. He said that I wasn't wanted at caucus night and that Andrew should warn me to stay home with my kids or there would be consequences. Andrew called me into his office to tell me about the conversation. He was in a tough position. As a first-term councilmember, he needed to maintain a good relationship with his colleagues. On the other hand, he felt protective of me as tensions escalated around the upcoming election.

"You should do what you want to do," he offered. "I am just telling you what he said."

Tears of disbelief at the veiled threat welled up in my eyes. But the shock and sadness didn't keep me from the caucus in the Cedar-Riverside neighborhood. Although I had canvassed for Noor when he ran for state senate, my role in the caucus was neutral. Weeks earlier, the chairman of the state senate district Democratic group asked me—the DFL district vice chair—to handle all the arrangements for the caucus and be site coordinator on the day. I was the only party official at the event who spoke both Somali and English.

Although sadly it would be far from the last, this was the first time I had received a direct threat concerning my political work. Despite a warning that I might get hurt, I actively chose to show up. Part of me didn't believe anything was going to happen to me, because fundamentally I'm an optimist. I couldn't imagine the people in Riverside, those closer to me than any other group in the world, physically attacking me. That didn't add up.

The other reason I insisted on attending the caucus was that if I, someone going against the accepted political discourse, backed down, then I would surely become a lasting symbol that Somalis can be intimidated and controlled. Unwilling to allow a future in which Somalis voted in lockstep for city and state politics, I decided that whatever came with attending the caucus was worth it.

By the time I entered the caucus at the Brian Coyle Community Center, I had already fielded many calls from the community center's staff and ordinary citizens letting me know that people in support of Noor had received death threats if they dared attend the event. This was on top of attempts to cancel registration for the event that night. I brought this new information to the attention of people higher up on the DFL and asked for extra security.

When a young man jumped on one of the tables and started shouting, I wasn't scared. The act seemed more ludicrous than anything else. There were rumors that these young kids, many of whom I had known since they were toddlers, had been promised free pizza if they caused a scene.

It was a different matter when one of the community elders, who was friendly with my father, began egging on the youth. I still wasn't afraid, but I was very angry. "If something happens to me," I said, although I still didn't really think it would, "I hope you will have a good answer for my father as to why you got all these people to hurt his daughter."

That was the last sentence I remember before I got hit. I didn't see it coming, because someone had pulled my hijab down over my face, for which I was grateful; otherwise I would have been covered in scratches. I tried to fight, but I couldn't see.

Pushed up against the wall, I felt blows from every angle and direction. Coming over and over, they almost felt unreal as I battled with the dark and noises all around. It was too much all at once. Someone hit my ear and a ringing sound filled my head.

Later, in videos the police showed me, I saw the confusion I felt. Nearly four hundred caucus-goers, people as intimate as family at a wedding, rushed to the front of the room—whether to throw a punch or block one, it was impossible to tell. Some might have wanted to hurt and others to help, but in the melee the two impulses were inextricable.

Wil, who was running Noor's campaign, pulled up to the community center just as all hell was breaking loose. Minneapolis police shut down the scene before delegates were selected, and people flooded out of the building. Noor, who had driven with Wil, ran into the building just as I ran out and hopped into the car's passenger seat.

"Oh my god!" Wil said, looking at my bleeding face, which was beginning to swell. "I have to get you to the hospital."

"First I need you to take a picture of my face and tweet it out—that I'm okay," I said.

"But . . ."

"I will not be bullied," I interrupted. "I want people

to see my face and know that I'm okay. I will show them that I am stronger than hate."

We did end up going to the hospital, where I was treated for a concussion, but that wasn't my first priority. I had worked to educate people in my community about their rights and power within our democratic system. The year before, I had cofounded the New Americans political action committee for exactly the purpose of bringing new immigrants into the political process.

When I stood up for Noor, I was standing up for every single Somali—not just on that day but every day forward. As I later wrote in an op-ed for the *Star Tribune*:

> I am a 31-year-old Somali-Muslim woman, a mother of three and an unapologetic progressive. Some suggest that as a woman, I meddle in political affairs and need to be "put in my place." Some say I deserved what I got because my opinions are contrary to those of a few male political leaders in our community. In addition, a small group has decided that one Somali elected official is enough and now the community should sit down and be quiet.

I would never be quiet, even if threatened with violence. If a Somali candidate wants to run for office of his own free will, no permission from the political establishment is required. What I have found in organizing and my political career is that anyone who believes gatekeepers can assure victory is always proven wrong. It isn't true just in the Somali community. No leader in a true democracy can promise that the election is in the bag.

Although I still have nerve pain in my neck from that night, it is nothing compared to the strength I take from the fact that the people who attacked me made a huge mistake. I know they wish they hadn't carried out their threats on me, because if they had done nothing, the system would have functioned as it had before. My standing up to their violence allowed me to make a much bigger statement than I ever could have on my own.

Your success and the successes of others you inspire can heal your wounds. Of all the wounds I have suffered, this is the one that is most healed, because every day I see the system I fought against get dismantled by people who used to feel so small but know now they too can be big.

# 15

# RUNNING

*2015–2016*
*Minneapolis, Minnesota*

We were in Noor's campaign headquarters anxiously waiting for the election results. In the beginning of his run against Kahn, I hadn't been as invested in this particular race so much as I was in trying to make democracy work for the people. After I was attacked, however, this election for me became truly about railing against the systems that served to intimidate voters in order to maintain the status quo.

I threw everything into the fight. I mean *everything.* I lost thirty pounds working my full-time job at City Hall, picking up my kids, heading with them over to the campaign, and doing whatever needed to be done while the children sat and did their homework.

This race was everything to me. When Noor came up short in the election and the entrenched political establishment that had gone so far to keep power prevailed, I was devastated.

The campaign headquarters were on the second floor of a two-story building that had a restaurant on the first level, where staff members and volunteers had started to gather. I came downstairs and realized that I couldn't be there.

Walking with Wil, who had run Noor's campaign, I saw my kids coming toward me. I was on the verge of tears. I didn't want to talk to or see anyone, especially my children. They were so much a part of this race, too. They probably wanted answers, and I didn't have any. So I ran. I ran from my children, bolting toward my car.

Wil was sprinting right behind me and jumped into my car.

"Let's just go!" I said.

We took turns screaming, cursing, crying, and calming each other down. I cried for thirty minutes—about failing others, giving people hope and ruining it. And then he took the next thirty-minute shift. We drove and drove and drove.

Once I had become involved, my heart was in it all the way. We needed change, and we needed it now. But we hadn't been strong enough to defeat them.

"Is that Wisconsin?" Wil asked.

We had sobbed so far down I-94 that we were almost out of the state.

"Oh yeah," I said. "We should probably go back."

**LIFE WENT ON, AS IT ALWAYS DOES, BUT I COULDN'T LET IT GO—** and neither could David Gilbert-Pederson. David—a longtime activist who at age seventeen went viral when he became the youngest delegate to the 2008 Democratic National Convention—and I had been part of each other's worlds for a long time. We had not only worked together on Andrew's campaign but were part of the kitchen cabinet for Mark Andrew, who had run for mayor in the same slate of progressive candidates that election season.

David grew up in Cedar-Riverside. This was our community. While others might have seen the race as regular politics, for us, it was personal. And we didn't feel Kahn represented us or considered our voices valuable.

David and I wanted to understand what went wrong with Noor's race and, ideally, figure out how we could recruit a candidate strong enough to withstand the shenanigans the opposition was willing to come up with. To that end, we started having lots of conversations with different members of our community, such as progressive friends who worked on other campaigns and community organizers from various causes. What we learned was that although people were unhappy with the incumbent, they weren't moved by the candidate himself. Many described feeling disconnected from him. "Noor's a nice guy, but I've never really worked with him," a young Democrat said.

"His credentials are great," a community organizer said, "but I don't remember him being here. Phyllis is at least the devil you know."

The other thing we kept hearing was "Ilhan, why don't you run?"

During this informal fact-finding mission, the question kept coming at me from different directions and in different forms.

"You have real community connections."

"If you were running, we wouldn't be invisible."

"I would knock on doors for you."

I couldn't imagine becoming a candidate for many reasons, the first being that I didn't know yet whether Noor would run again in 2016. I was also the mother of three children, one of whom was still very young. I would have to figure out childcare and getting home by dinner, things that many other women running for office worry about. On top of that, I had to consider what it would mean to the Somali community, who had never had a female run, let alone govern. Would just entering the race humiliate every male in my family?

Perhaps my greatest hesitation, however, was the political establishment itself. If I was as good a candidate as folks were saying, what would those in power be willing to do to keep me from winning? They were willing to come at me with physical violence for merely supporting the opposition. I couldn't imagine the lengths they might go to if I was the one on the ballot.

My female colleagues weren't having it. "You make the ask of other women all the time. You tell them they have to be brave enough to fight for the things they want," my friend Pakou said. "How can you be a hypocrite and have your own reservations?"

"But you don't understand," I said.

Uncomfortable with and unused to being accused of cowardice, I eventually realized all my excuses were just that. I didn't want to make my life harder on myself. But, I reasoned, my life would become hard anyway if I had to be represented by a person and a system I found totally unacceptable.

The last hurdle was the question of whether enough voters would support someone as strange as me.

**IN OUR HOUSE DISTRICT, THERE WERE THREE RELIABLE VOTING** blocs—seniors, Somalis, and students. Kahn and her team affectionately called these groups the three S's.

I was expected to mobilize a new generation of voters and excite new immigrants, but the people I really needed to win over were those who had voted for Kahn for the last forty-four years. To be a viable candidate, you have to shift the longtime residents, which had never really happened.

In the past, candidates like Noor would get a percentage of whoever the new American population at the time was—Koreans, Vietnamese, then Africans—and maybe a few longtime residents. But that was no match for Kahn's base. "You don't know this new person," she said, scaring voters again and again. "They're not really from here."

David, who as I said couldn't let it go, suggested I meet his mother as a focus group of one. Although his birth parents were Filipino, David was raised by white adoptive parents.

"My mom has lived here forever, but she doesn't really get excited about politics or politicians," he said. "We should have a conversation with her about what it would mean for somebody like you to run for office."

We met with David's mom early in the morning in our favorite place, the local Perkins coffee shop, where we would go to do our political plotting. Perkins is paradise for coffee addicts like myself—for one reasonable price you can buy yourself a free flow of the stuff for one hour or six. Open twenty-four hours a day, it's the kind of place that encourages lingering, which is particularly dangerous for me. Many people refer to me as the chair-breaker, because when I'm comfortable, I can sit somewhere for hours. And when I have coffee, I'm comfortable. I'm not the only one at home at Perkins. In fact their coffee fuels all kinds of political operatives from early morning meetings to late-night wrap-up conversations to middle-of-the-night delirious brainstorming sessions. I was so attached to the last one in Minneapolis that when it shut down I actively considered renting the vacant building for our campaign office.

Over steaming cups of coffee with David's mother, we talked about what mattered to her. Although she fit the demographic of Kahn's base to a tee, she had voted for Noor.

Then I shared my story with her, telling her what mattered to me. At some point, she started crying. David and I looked at each other nervously.

"I never felt inspired by anyone who wanted to be in office in our community or our state and country," she said. "I know you're not really ready to say you're running. But if you did, I will not only vote for you but I will donate. And I have never done that before."

David was right. His mom was the fuel we needed. Early one morning she set up a meet-and-greet in her home, with four people. That led to another meeting with about six people on a Sunday afternoon, at another person's home. The informal gatherings quickly grew to ten, then twenty. When thirty people showed up, David and I didn't know what to do with ourselves. I started to cry. Those numbers may not seem big, but in a district of around 40,000 voters, the rapid growth over a three-month period was tremendous.

By the time we formally kicked off my campaign on October 10, 2015, we already had in place our slogan, "Time for Ilhan" (created by my friends Jillia and Lena); pledges that added up to $10,000 in donations (although by the end of that night another $3,000 came in, for an astonishing total of $13,000); surrogates in every single neighborhood supporting my candidacy; and a future we were excited to build.

It was a good solid start. Maybe a little too good, because no sooner did I announce that I was running than I became a threat—and the infighting started.

About two weeks after my announcement, the elders of the Somali community summoned my father to a meeting, where they announced their support for Noor, who had entered the race shortly after I announced, and told my dad to ask me to withdraw.

"She believes she can do this. I believe she can do this," Aabe argued. "Noor had the opportunity, with my daughter assisting him, and he wasn't able to win. It's unfortunate that instead of saying they can both run or asking Noor not to run, you're asking her. Especially when she has all of this momentum. People are so excited about her running."

"It's not appropriate for a woman to lead," one of them responded.

"You should at least talk face-to-face with her," Aabe insisted. "I see Noor waiting outside to meet with you. I can have her here in a second. They can meet with you together."

"It is beneath the elders to have a girl sit across from us," said another, before explaining that they planned to use all their resources to fight me if I insisted on continuing to run. My father should also not consider returning to complain later when I was shunned.

"This is your warning," the elder said. "There will be no other."

"When she wins," my father replied, "I hope you know that I will tell her not to take any meetings with you."

They laughed. My dad left, shattered for the first time to have been a part of that community. I could see the hurt when he relayed the details of the meeting to me later. "I don't even understand why you went," I said to him. "This isn't a process for the elders to negotiate."

This was an exercise in the democracy in which Noor and I promised each other that whoever had more delegates in the convention, the other would drop out. It was a typical pledge among political candidates. Noor's promise was more than enough for me.

The elders kept their word and a terrible smear campaign spread through the Somali community. Disgusting videos of me began to pop up on social media. There were the classic sexist ones that portrayed me as brainless and controlled by sex, money, or drugs. The main theme was the backward idea that women can't want things for ourselves or that we couldn't possibly have the intelligence to succeed where men have failed. There had to be something else, something more sinister.

I was the Manchurian Candidate. The made-up theories about me from that time are why none of the bizarre conspiracies concocted by right-wing groups bother me now. The idea that I'm a Qatari plant in the government is way less creepy than the Somali videos where I'm talking but instead of my voice, devilish male tones come out of my mouth.

Facebook was flooded with these clips of me giving speeches—with a male voice—about the same clan conflicts that tore apart my native country. Because Noor and I are from the same clan, the male possessing me was from my husband's clan, and his message was that they were trying to destroy us.

"We killed you and chased you out of Somalia," the voice said as my lips moved. "Now we are going to make sure Noor doesn't win, and you don't have any success in America."

This message opened up a massive wound that still festered among the clans, even in Minnesota. The old divisions were employed throughout the race, at every level of it. None of my extended family members wanted to be around me because of the hell I had unleashed. My poor dad was bullied at the mosque.

But I didn't have time to become indignant or dispirited; I was too busy trying to run a campaign. I was out all the time, talking to anyone who would meet with me from students at the University's LGBTQ center to a senior Somali yelling at me in her kitchen.

**DESPITE MY DAD'S EXPERIENCE IN MEETING WITH THE ELDERS, I** didn't give up on trying to win over influencers in the Somali community. I believe that when you want to change people's minds, the only way to do it is one-on-one, human to human.

Ali Ganey was one of those influencers. Politics was in his blood. Although he never had a formal role, he had been organizing since the Utange camp, where he was in one of the first groups of Somalis to leave for the United States. In Minneapolis, he had established himself as a successful entrepreneur and was known for providing financial and networking support to candidates back in Somalia.

We knew of each other but had never exchanged more than a hello. I found him on Facebook and sent him a message, asking if he would have a coffee with me.

A couple of weeks before the convention, we met at a coffee shop called Caribou, where he began by laughing at me. "Why are you running?" he asked. "Don't you have a good job?"

In what I would learn was characteristic Ali, he was having a really good time treating me like I was insane.

He had seen the dirty videos on Facebook and wondered why I would expose myself to those kinds of attacks. He thought I—a woman from a patriarchal society with no experience in elected office—was wasting my time.

"You are crazy, girl," he said. "You should stay away from this."

"Explain to me how my bettering the community that I have to raise my children in is wasting my time," I said, "while you spend hours every day on people running for office in Somalia?"

"Well, because they have a chance."

"You didn't even ask me about my chance."

Then I showed him my numbers: the precincts I was able to pull together, the money I had raised, the volunteers working on my campaign day and night. Ali likes numbers, but he wasn't wholly convinced. "I'm more Republican than progressive liberal," he said.

I had made my case to him, which was all I could do.

"I'm running to improve life for a district, not just the Somali community," I said. "Whether they come out to support me or not, I am determined to be their representative."

# 16

# YOU GET WHAT YOU ORGANIZE FOR

*2016*

*Minneapolis, Minnesota*

The convention, on April 9, 2016, to determine who would win the Democratic Party's endorsement, was taking forever. Noor's and Kahn's campaigns were trying everything in the book to mess with mine: from questioning credentials to slowing down the ballot process. They might have hoped that my delegates, the majority of whom were young people, wouldn't be able to sit through what turned out to be a twelve-hour convention on a Saturday. But they did.

On the first ballot, I led with 55 percent of the delegates to Kahn's 35 percent and Noor's 11 percent. After the second ballot, it was clear Noor didn't have enough delegates to win. Despite his earlier promise to me, however, he didn't drop out. Not only that, he didn't leave the convention—which meant his delegates were all also going to stay.

I walked straight up to him, leaned down close, and in Somali, reminded him of his pledge of unification—that we would not allow our community to be torn apart for public theater.

He wouldn't even look me in the eye. I believe it is because he knew everything I was saying was true. That moment, as I suspected at the time it would be, was his undoing. Nothing could erase that image, even two years later, as Noor ran for my seat when I was running for Congress. The gender-based attacks he had deployed against me were turned on him as Somalis openly mocked him about that day.

"You were so afraid that a girl would beat you."

"I'm so glad she finally gave you the seat that you cried for, for years."

Although by the end of the convention, none of us three candidates was able to claim the party's endorsement, which required 60 percent of the delegates, I told my father, "This is probably the best thing that has happened to us. It's like a walking billboard for the men in our community not to mess with us."

**NOT THAT THEY DIDN'T TRY. FROM CLAIMS THAT MY CHILDREN** weren't my own because I was infertile to the idea that all children would get autism if I were elected—because I had a mandatory vaccination bill in the works—the rumors never stopped.

Ali called me early in the morning, nearly every day, to say, "A bomb dropped today."

After the convention debacle, Ali went from a supporter in name only to one of my biggest champions. As he put it, "Taking advantage of the American system to block this woman, who worked hard and deserves to win, because of cultural bullshit—now it's personal to me." He wasn't kidding; Ali had cried after the fifth ballot recount.

From that day on, he approached any elder who would hear him for *garcelis*, the Somali practice of debate over coffee. At a Somali community event, he gave a speech so impassioned that at one point he started to cry. "If you don't give Ilhan money or help her win this campaign, then I'm just going to sell my children," Ali said, which was a Somali way of saying, "I'm willing to die for this person," since children are worth more than your own life.

Every rumor or insinuation the opposition dropped about me drove him crazy. Perhaps the silliest fabrication came out of my attending a vigil held in June 2016 by the LGBTQ community in the aftermath of the mass shooting at Pulse, a gay nightclub in Orlando. Perpetrated by a man who pledged allegiance to ISIS, it was the deadliest mass shooting in U.S. history at the time.

Standing with my friends, Erin Maye Quade, who was also running for a seat on the state legislature, and Phillipe Cunningham, a transgender member of the city council who at the time worked for the mayor of the city, I was exhausted. It was Ramadan, and I was fasting. But that wasn't the source of my weariness, because Erin and Phillipe felt the same way. One mass shooting in our country followed another, and none of them made sense.

As the three of us held one another and wept, a host of candidates for office took to the mic to speak. Everybody wanted to say something, as everybody should speak out against such hatred. But unfortunately it was election time, so people were taking forever to talk. Kahn enumerated her entire LGBTQ record over her decades in office. To me, it didn't feel right to be campaigning at a vigil.

So when the moderator approached Erin and me to say that she would speak first and then it would be my turn, I turned to Erin and said, "I don't know how you feel, but I'm not in the mood to say anything. I am here just to be with community."

"Me, too!" Erin said. "I've been thinking about that this whole time. I don't want to decline the offer, either, because that's weird, too. What should we do?"

"What if we just go up together? And just say something short like 'I'm Muslim, and a candidate, and I'm queer, and a candidate. All we want are to be friends and have safe communities to live in.' "

Erin liked the plan, so we got onstage and, still holding each other's hands, each said our short piece. It was a small, meaningful moment in the face of a meaningless horror.

Some Somalis and right-wing groups got ahold of the image of Erin and me and ran with a story that our holding each other's hands close to our chests was our secret lesbian lovers' pledge. Memes exploded using hateful words like *faggot* in both English and Somali. Although I was deeply offended by the vulgarity, I didn't want to address the issue for fear of validating the perpetrators in any way. Instead, I made that image the home page of my election website.

**THOSE WERE THE KINDS OF BOMBSHELLS THAT DISTRESSED** not only Ali but my entire campaign staff. I walked into the office in the morning more often than not to find everybody huddled in front of a computer. They were chewing their nails while reading the latest dumb thing some random person had come up with—instead of working on the campaign strategy for the day. Finally I had to put an end to the diversions.

"We have to stop," I said. "I don't want a single conversation about whatever the latest is. They're doing this precisely to occupy our brains and our time. If we're constantly reacting to their insane stuff, we're not focused on what we need to do to win. None of this means anything. It is just for the purpose of distracting you."

I went on, "Remember when I said, 'I don't know how people are going to deal with this extremely unique person running'? This is what I meant. Anything and everything can be made up about me. I represent so many different communities—immigrant, female, person of color, Muslim, Somali—each with its own stereotypes to overcome. Those against me would have me own every single negative label ever attributed to any one of my groups."

"You have an image we have to protect," one of my staffers said.

"I am in the image of everyone that looks like me and has my voice. There is no use in trying to get ahead of it. There is no messaging, there's no defending, there's no conditioning. I just have to be comfortable knowing who I am and what I stand for. And you just have to do the work, which is to build enough familiarity with people so that when they see lies about me, they'll think, *That's crazy.*"

The speech was more of a pep talk than anything else, since my staff—which other than East African volunteers was predominantly made up of students, most of them under the age of twenty-five and first-time voters—worked harder for my campaign than I would have ever imagined possible.

David, who became my campaign chair; my campaign manager, Dan Cox; and his deputy and University of Minnesota student body president, Joelle Stangler, simply refused to believe that what we were doing was impossible. As much as they believed, however, they ran and reran the numbers, double-checking their data, making sure we knocked on every single door.

In the dead heat of late summer, as the August primary neared, we did so much canvassing not just because we had to return to confirm the support of voters whom we had convinced. We continued to go back day after day to knock on the doors of those we had not convinced. Six, seven, eight hours a day. From morning to night. Upstairs and down, walking miles through neighborhoods. To this day, my sides hurt when I think about that race.

No matter how exhausting it was, a blitz on the doors was exactly what I wanted to be doing. Face-to-face encounters didn't just form my campaign strategy. They were also how I honed my platform to be one that truly represented *all* the district's residents—even those who hadn't voted in the past. While I was always committed to affordable education, income equality, criminal justice reform, climate protection, and support for new Americans, I talked to hundreds of constituents to figure out how to best advocate for those broad issues within the context of this community.

I was at the doors all the time, because I was panicked about not giving the race my all. If I lost, I wanted it to be because people didn't want to vote for me, not because I hadn't done everything within my power to win. The campaign was hard but also rewarding. It was the most painful and joyous thing I've ever done outside of giving birth. Every day a beautiful thing would happen, even on the same social media platforms from which hatred against me spewed.

**SOMALIS AROUND THE GLOBE HAD WATCHED THE APRIL CON-**vention, which had been streamed on Facebook, and unlike the Somali establishment in Minneapolis, they were cheering me on. They looked to promote me in any way, so that I couldn't even sneeze without them sharing it enthusiastically. My online armor, they drowned out my harshest critics with their overwhelming support for me to keep on going.

The most beautiful moments, though, were with Aabe. I took him with me for an intimate meeting with a running club. Up until that point, I'm not sure anything about my running for office was real to him, except his unpleasant meeting with the Somali elders. It had left such an indelible impression on him that he believed everyone had an equally harsh opinion of me. So when the friends in this running club gushed over me in a supporter's living room, I could see how happy he was. "You should be so proud of what she's been able to accomplish," a woman told him. "She's a wonderful kid!"

Aabe's love for me was never in doubt, but when my press spokesperson, Jean Heyer, asked him to do one of my fundraising surrogate letters, I saw it expressed in a new way.

He had been reluctant to do the letter. "I don't want to do interviews," he said. "What if I say the wrong thing, and she writes it down?" I explained this wasn't for the press but was an email about what he thought of my running for the state legislature. Then he said yes. Still, when we sat down with Jean, I could tell he was nervous. She started with a softball question. "What do you like about Ilhan?"

Aabe answered as honestly and directly as if he were being deposed. "I like everything about her," he said.

Tears ran down my face at my father's genuine feeling. Jean ended up titling his email "I Like Everything About Ilhan."

**I BELIEVE YOU GET WHAT YOU ORGANIZE FOR, BUT ON THE DAY** of the August 9 house primaries, it was completely unclear who would win. I trusted our work. And no matter the outcome, I was going to be content because I knew we had done everything we had set out to do. We were smarter, more organized, more energetic, more everything than our opponents. All the massive stones they threw at us served only to make us tougher. That didn't keep them from heaving boulders up until the bitter end.

But you don't know until you know. After the polls closed, I didn't want to go back to any of the polling places or the viewing areas. I also didn't want to go to our war room to wait for the results to come in. So I did what I always do when I need to calm my nerves; I went driving.

I was headed down my familiar and soothing route along I-94 when Joelle called. "They're almost done," she said. "You should get back."

In the war room, all my anxiety returned. The numbers weren't in. I ran back and forth, aggressively trying to find out just what the holdup with the results was about. And then the race was called on Twitter. I wanted to believe the social media post's results, but I continued to refresh the website of the Minnesota secretary of state relentlessly until the official results popped up.

I collapsed to the floor as the events of the last six months hit me all at once. All of it. All the emotions. All the stress. All the tiredness. All the doors slammed in our faces and the ones open wide. All the debates online and in person. All the helping hands and beautiful faces. All of it rushed through my head like scenes in a movie. I guess it felt unreal—I couldn't believe it was done. We had won.

# 17

# AMERICA'S HOPE AND THE PRESIDENT'S NIGHTMARE

*2016–2017*
*Minneapolis, Minnesota*

The story in the community was that a Somali blog was paid to put up the post until it got picked up by an English-speaking site. Then, just as suddenly and anonymously as it had appeared, the original blog source disappeared.

The post established two simple facts, which were

true, but that unleashed a torrent of speculation and conspiracies that follow me to this day. They were that I had applied for a marriage license in 2002 and then again, in 2009, when I eloped.

These facts—which are true—were couched in speculation that I had married a relative illegally, to get him entry into the United States—which isn't true. The blog post appeared on the Somali Spot, an online Somali discussion forum, and was brought to my attention on August 5, 2016, the weekend before my primary win. It was the epitome of a last-gasp effort to keep me from winning the district seat. If someone had accurate information or true concerns that I had committed an immigration crime, they would have gone to the *Star Tribune* with it. Not the Somali Spot.

Unsurprisingly, the news had no impact. That is, until a few days later, when a conservative website got ahold of it, most likely thanks to the same sources who had planted the original post to begin with. From there, it was picked up by the *Star Tribune* and then the whole country.

This all happened *after* my primary win, when the opposition knew for a fact that my winning the general election in the firmly Democratic district was a foregone conclusion. Instead of celebrating the first black Muslim woman elected to a state legislature in the United States, some people in my own community continued to work against it.

That Somalis were some of my harshest critics might

seem absurd. But they refused to accept me because I wouldn't kiss the ring. It goes back to my inability since childhood to submit to bullies. The source of the information about my brief past relationship was the work of people trying to preserve their power structure. They might have been able to manipulate Noor, but they knew they couldn't do the same with me. So they kept trying to teach me a lesson.

It didn't work. Despite all their efforts, I kept rising. Ironically, I don't know if I would have the widespread support I do today if I hadn't been messed with as much as I have. Influence and enemies go hand in hand.

**I LEARNED THIS LESSON EXPLICITLY FROM NANCY PELOSI, THE** first woman to become Speaker of the United States House of Representatives and someone I admire greatly. In the summer of 2019, she invited me on a congressional delegation to Ghana. During the flight, she told me about her journey: from her first campaign in 1987 when she was elected to Congress to becoming the highest-ranking female elected official in United States history, not once but twice. Then she turned the conversation to me.

"It is really moving to see the number of people who come out to support you," she said.

"Yes, but every time I'm attacked, the attack seems to be amplified by the support," I replied, confiding in her because it might make more sense to her than to most.

"Sometimes I wish I could shoo them all away and take the beating myself so that I don't continue to feed the craziness."

She sighed and looked at me with a maternal gaze, one that had a lot of knowledge behind it. Nancy is the kind of person who, by the time we landed, knew Ghana's economy down to the percentage change in its cocoa bean exports over the last year but who also put a blanket around me when I fell asleep in my seat.

"The amount of money that is spent on ads and anything else to try to take me down is mind-boggling," she said. "The truth is, it is a badge of honor to have this many people invested in one's failure. If they weren't afraid of your power, they wouldn't work so hard to erode it."

To hear that message from the powerful being she is meant more than words can express. But this conversation happened much later.

In the days after my primary win, I couldn't comprehend why a former romantic relationship of mine, long over, was such big news. The creepiest part of the whole thing was that as soon as the conspiracy theory about my marrying my brother went viral, the original post disappeared. It vanished in a way that is not supposed to be able to happen on the internet. To this day, I can't understand how they managed it. But the various outlets that had picked up the story reported that the original post had disappeared, reinforcing the conspiratorial narrative that I had the powers of a foreign villain.

I was catapulted back to the *Star Tribune*'s boardroom when I had sat next to Andrew Johnson as he faced a litany of questions about the angst-ridden blog he had written as a teenager. This is the kind of personal stuff that gets dredged up in a political career. Although it has absolutely no practical or moral bearing on a candidate's viability to hold office, one has to deal with it.

The people who planted the story hoped that I was a thin-skinned woman who wouldn't have the fortitude to deal with the scrutiny or pressure that came with the media storm that ensued. As I told the host of lawyers and political operatives sent to my aid, "I'm fine. But I didn't do anything wrong, and I don't want to talk to anybody." Once they understood that I hadn't broken any laws, they disappeared.

**THE PERSONAL TOLL OF THESE FABRICATIONS WAS A DIFFERENT** matter, however. My kids, thank god, were shielded from the drama thanks to their young age and the fact that they were away on a mini vacation for the height of the news cycle. I wish I could have protected my entire family the same way.

My relatives, who had honored my space and made peace with the messy, painful process I had needed to go through to become who I wanted to be, had no interest in my reliving it. The humiliation and pressure wasn't directed just at me, but at every single person who had invested in my election and my life.

When the man who prays with my father every single morning saw a meme on social media that I would be going to jail for forty years, he felt it was his duty to warn Aabe of my imminent peril.

"She should think of her children!" he said. "They are going to lock her up."

"For what?" my father replied.

It might seem silly to outsiders, but many immigrants in my district came from countries where leaders *do* lock people up for challenging the status quo. This was what the Somali establishment—who had been against my election—counted on: that the level of obsession and panic in the community would be too much for my family. Just as in my case, they hadn't foreseen that my father and my husband would be as resilient as they are.

Still, the event created a huge strain on my relationship with my husband. The rumor was meant to humiliate him, and it did.

"You must be really embarrassed," someone told him.

"If she just resigned, this would all go away," another said.

It was maddening for my family, not only because people were lying in disturbing ways about someone they loved but also because they weren't allowed to have any kind of reaction. You can't get angry if you are associated with a representative. If my father were to shout in the mosque or if my husband threw a punch, it would instantly go viral. I was no longer just theirs.

Aabe and Ahmed worried about being liabilities. Not everyone can take everything all the time. Sometimes it's easier to give up on the association. The level of worry isn't good for them, and it isn't good for me to see that level of worry on their faces, either. Space builds as people distance themselves from you, or you distance yourself from them, because the dynamic is just too painful.

I've experienced this slow-breaking-away effect with almost all of my family and closest friends. The media and public scrutiny as my profile grew were particularly hard when it came to anything related to my marriage. That's because I've always been a very private person. I never felt comfortable holding hands or even saying the words *my boyfriend* out loud.

However, anyone who runs for elected office has to talk about themselves. A lot. That alone was enough of a stripping for me. The personalized aspect of being a candidate, however, left me feeling completely naked.

Other than when I was pregnant for the first time or when I was unraveling and left for North Dakota, I didn't invite people into my private life. As the friend I confided in before I left for North Dakota said, he hadn't even known what I ate for breakfast before my breakdown. I'm not a person who lives in feelings. Being open isn't my natural state; working is. I am never comfortable talking about my relationships.

I wound up issuing a statement about the insinuations that had first appeared on the Somali Spot, explaining the particulars of my marriage ceremonies and separations, both civil and religious. "Like all families, we have had our ups and downs," the statement read, "but we are proud to have come through it together."

In retrospect, though, I wish I hadn't given even the basics of the how and why. Not only did the speculation on my romantic past play into a xenophobic, anti-immigrant mindset, but it wasn't effective in putting an end to the intrigue.

In a world of conspiracy theories and propaganda, which flourished with the rise of Donald Trump, no answer was ever going to be good enough. Indeed, in 2019, President Trump repeated the baseless conspiracy theory that I had committed immigration fraud, nearly three years after I thought I had answered it once and for all.

**WHAT IS THERE TO SAY ABOUT THE ELECTION OF DONALD** Trump as the forty-fifth president of the United States? It was tragic.

This was a man who at a campaign rally held two days before the 2016 presidential election at the Minneapolis–St. Paul airport singled out Somali immigrants as radicals who shouldn't have been let into the country. "A Trump administration will not admit any refugees without the support of the local communities where they are being placed," he told thousands of supporters who showed up to the event. "It's the least they could do for you. You've suffered enough in Minnesota."

For months we had knocked on doors and held massive rallies to get out the vote for his Democratic opponent, Hillary Clinton. I traveled across the state, sharing the message that we in Minnesota "don't get mad." Instead, I implored people to respond to Trump by voting—in record numbers. A lot more immigrants voted in that election cycle than ever before, but it wasn't enough.

Clinton won Minnesota—but by less than 50,000 votes, just 1.5 percentage points ahead of Trump, compared to Barack Obama's nearly 8-point margin in 2012. This was a state a Republican presidential candidate hadn't carried since Richard Nixon!

Just as I was supposed to be celebrating my victory as the first Somali American lawmaker in the United States, I was grappling with how Trump's hateful divisive message had resonated with our neighbors. In the car ride home on election night, after the results had come in, I admitted to my team it was "scary that his hateful rhetoric can find a partner in the hearts of so many in our communities. How different the world of tomorrow is going to be for many of us."

The fight-or-flight instinct that was already overdeveloped in my brain flared. The only antidote was deconstructing his success, which took some time. I still spend time analyzing it and have come to the conclusion that there is no one answer.

There are many explanations that make sense. Clinton wasn't the best candidate. Economics also played a part. Many farming and labor communities in Minnesota, desperate for solutions to the lack of opportunity in a rapidly changing world, handed their precincts to Trump. And then there was his signature style that contributed to his success. As Americans, we think of ourselves as bold and brave. The vulgarity of Trump's character is appealing to people for whom it doesn't feel very American to speak in politically correct terms or conform to the rigors of empathy or subtlety. You tell people what they should think. Not the other way around.

In a game of political chess, some ultra-liberals also voted for Trump. There was a contingent that thought he was too stupid and corrupt to inflict as much of what they considered damage as the neoliberals who would have advanced with Clinton could. Antiwar, anti–status quo, these voters thought that by the time Trump figured out how to move the smallest piece of his absurd agenda, it would be too late.

They were all in for a rude awakening. Trump, whose sole motivation is his own self-interest, was willing to hand over the reins to anyone as long as they gave him what he wanted. And it didn't take him very long at all to do some very real damage.

**ON JANUARY 27, 2017, TWENTY-FOUR DAYS AFTER THE START OF** my first term in the Minnesota legislature, Trump signed an executive order to block and ban visitors, refugees, and immigrants from seven predominantly Muslim countries, including Somalia. Although the order was euphemistically called a "travel ban," the president had made his real intent perfectly clear as a candidate two years earlier. During his campaign, he had called for "a total and complete shutdown of Muslims entering the United States until our country's representatives can figure out what the hell is going on."

The effect of the ban on my little district was chaos. Questions and concerns immediately flooded my office. People were scared. We had to do something. A few hours after the ban was announced, I held a press conference, during which I invited the president to tea so I could educate him on the people he seemed to fear so much. I also sent out a mass email for an event at the district office. Although it would be small, it was important to get the community together. Then I took a nap.

When I woke up an hour later, thirteen thousand people had RSVP'd to the event. And there were seventy-seven messages on my phone—the majority of them from our police and fire chiefs about the codes I was about to break in the small room that was my district office. (We ended up moving the event to the Brian Coyle Community Center, where over a thousand people showed up.)

It was already becoming apparent that I was being held up not just as a public servant representing those directly under attack by the current administration but as a counternarrative to Trump.

In Minnesota, the Democratic Party was demoralized after losing its majorities in both houses. The Republican caucus won one of its largest house majorities and a narrow but unexpected flip in the chamber. Then here comes Ilhan Omar, a black Muslim Somali refugee turned American legislator, to give Minnesota Democrats pride. I became a symbol and a message, not just in Minnesota but also around the world.

The number of requests that came into my office from day one was unprecedented. Global travel, speaking engagements, interviews: they stacked up. Media outlets from around the world, which tracked our press releases and my movements, constantly asked for comment. Every day, hundreds of requests came in, so many that we barely had time to go through them all. "This is like nothing I've ever seen," said my legislative assistant, Connor McNutt.

I first met Connor on a freshman class retreat in a suburb of Minneapolis, where he was one of the statehouse staff members on hand to help us learn about and get acclimated to our new positions. Even after an all-day training session, when we started talking about politics, we didn't stop for an hour and a half. We connected in part because Connor grew up in Faribault, Minnesota, where there is a sizable population of Somali immigrants among its population of roughly 24,000. He was not unfamiliar with the stigma attached to new immigrant populations, and the challenges of mine in particular.

"Maybe they'll add you to my list," the affable Connor joked, since he already represented two of the busiest members of the Minnesota house at the time.

"That'd be a lot," I laughed.

That's exactly what it was—and more—when five days before my swearing-in, his supervisor informed him that I was being added to his roster. Connor had been looking for someone to lighten his load a bit. Instead he got me. When he called to check in and get up to speed, I was already crazy busy with everything from a swearing-in ceremony on the Quran with more than a hundred guests to a cover shoot for *Time* magazine.

**THE SEPTEMBER 18, 2017, COVER OF *TIME*, WHICH FEATURED MY** picture, made international news—as did my line "I am America's hope and the president's nightmare," when I appeared on *The Daily Show with Trevor Noah* in July. But the moment I realized that my profile had become outsized for my position had come months earlier, before I had even taken an oath of office.

I was in Washington, D.C., during the first week of December 2016 for a day of policy training geared toward state legislators at the White House. President Obama's longtime advisor, Valerie Jarrett, led the event, where we learned about various ideas we could implement in our home states.

On the cab ride back to my hotel, I suggested another route to the driver, who flew into a rage.

"Shut up," he said.

"Are you talking to me?" I asked in shock.

"Yes," he said. "Do you think I want to chauffeur around filthy people like you?"

His rant escalated—along with my anxiety—until we arrived and he finished with, "I should remove your headscarf and see what you have hidden under it, you ISIS."

I knew there had been a rise in hate crimes against Muslims. While more than half the incidents motivated by religion targeted Jewish people, the FBI reported that in 2015 there was a 67 percent increase in crimes against Muslims. Facts and figures didn't do anything to dampen the alarm I felt being on the receiving end of such hatred. I took to Facebook to post about how I had been "subjected to the most hateful, derogatory, Islamophobic, sexist taunts and threats I have ever experienced."

In an uncharacteristic moment of vulnerability, I wrote: "I am still shaken by this incident and can't wrap my head around how bold [people] are becoming in displaying their hate toward Muslims. I pray for his humanity and for all those who harbor hate in their hearts."

I thought I was communicating with my friends and about 10,000 followers. What I didn't realize was that in the month between my election and that day, I now had over 160,000 followers. Shared over a thousand times, the post instantly went viral and wound up on the news sites of every single outlet in the United States and of many others abroad. I later appeared on *The Rachel Maddow Show* to address the controversy.

In the moment that I posted about the incident, however, I had not yet made the transition to my new reality. It was jarring to finish a campaign for state rep one day and to meet the unrealistic expectation of standing up to a demagogue president the next. The widespread interest surprised me and, frankly, continues to surprise me. Notoriety is something I will never internalize.

As much as Americans on one side of the aisle thought I was some kind of hero, there were those on the other who believed I was the devil. Just as my office was inundated with appearance and interview requests from our first day, we also received countless hate calls, mail, and emails.

**WHILE PEOPLE FROM ALL OVER THE COUNTRY REACHED OUT TO** call me disgusting things every day, they really ramped up after moments like my press conference on the Muslim ban. As the face of the opposition, any time I appeared on the national stage there was a significant uptick to the tune of several dozen death threats in a twenty-four-hour period.

All the threats on my life came directly to Connor, who worked with our sergeant at arms at the Minnesota house as well as with the Minneapolis Police Department to up patrols around my home. They did everything they could, but resources were limited. There simply wasn't the capacity to find the source of every threat on an elected official and follow up to see if the threat was at all credible.

Later, when I went to Congress, my office was assigned a Capitol Police special agent in the Threats Division, who works closely with the FBI on everything from monitoring online chat rooms to doing background checks on and eventually visiting the individuals who make the threats. Mostly the perpetrators are people whipped into a frenzy by right-wing television or online news hosts, for whom I'm a punching bag. They have no real means by which to carry out their threats. In the rare case that weapons or a background of violent behavior is discovered, the agents turn the case over to the appropriate district attorney to press charges.

In Minnesota, it was pretty much just the sergeant at arms and Connor, who would get very concerned by all the hate mail. At first he worried about alarming me, but he quickly learned I'd rather know when a letter with a bull's-eye with my face at the center arrives at the office.

Connor was more than nervous; he was also upset at the new normal. As he said, "It isn't right that people feel empowered to say these things, that humans think it's okay to do this to another human being."

It is hardest on my staff when there's a rise in the verbal attacks on me, because they are the ones who have to answer the phones. We don't have the luxury of not answering the phones, since even if it's one out of ten, there are people who call because they need help. Unfortunately, the task often falls to our interns, the youngest staffers among an already young group. But we prep them for it, mainly by telling them to take breaks whenever they need them and to get off the phone as quickly as possible when they do receive a hate call. "You don't have to sit and listen to that garbage," Connor told an intern. "Just say, 'Thank you, have a nice day,' and hang up."

**WE MINNESOTANS HAVE A HARD TIME BEING MEAN AND DON'T** fare well in the face of hate. In the summer of 2019, in a rally in Greenville, North Carolina, President Trump accused me of being anti-Semitic, supporting Al-Qaeda, and looking down on "hardworking Americans." His accusations were met with nearly eight thousand supporters chanting, "Send her back!" My friends from home were horrified.

"There's a Trump rally and there is chanting," Jean

texted me. "I feel sick to my stomach. Text me back and let me know you are safe."

Wil later told me he had wept while watching the rally on TV.

When my phone started to blow up with texts and calls, I hadn't yet seen the rally because I was coming from the Capitol and was late for a dinner with a group that included Bernie Sanders. When I arrived at the restaurant, Bernie had saved a seat for me next to him at the table. He took my hand when I sat down and asked, "Are you okay?"

I assured him that I was.

"I've heard that all my life, out of cars, in grocery stores," I said. "I survived war. I'm not afraid of these people. What I worry about are the people who look like me."

I was a member of Congress. The idea that he or anyone else could "send me back" was ludicrous. Much more real, however, were the immigrants, Muslims, and anyone else who felt an automatic association to me because of my identities taking the televised moment as a powerful symbol of rejection.

They weren't just chanting, but chanting in front of the president of the United States, whose expression was one of obvious appreciation. That is the ultimate affirmation of the rejection many of us already suspect. *They don't want me here and my president agrees.*

I remember when I graduated from eighth grade, I received a presidential certificate of excellence signed by Bill Clinton. It didn't matter how many other kids got the same piece of paper; Aabe and I basked in the pride that the president had affirmed *my* hard work. Politics aside, the highest office in the land carries great weight.

**BY THE TIME THE GREENVILLE RALLY HAPPENED, IT HAD BEEN A** long time since I'd worried about myself. Years earlier, right after the president introduced the Muslim ban, I did get scared. I was about to travel and feared for my own safety. Aabe was incredulous when I confided my anxieties in him.

"What are you talking about?" he said. "You have a responsibility to be strong for everyone else who doesn't have the power and the influence that you have. It would be wrong for you to worry about you. Your job is to worry about everyone else."

That was the reorientation I needed. Yes, hate affects me. But I represent constituents who don't have the ability or resources to defend themselves against the assaults on their lives, whether that's the practice of separating children from parents seeking asylum at the border or of dismantling environmental protections in an already warming world.

From the Muslim ban onward, I had the painful realization that even though most of what the administration was going to throw at us was very personal in that their proposals would have an actual impact on me and those that I love, I couldn't grieve because I had to show up for my constituents and the country. At least when I was fighting for them, I was no longer afraid.

# 18

# WALKING IN LIKE A WHITE MAN

*2017–2018*
*Minneapolis, Minnesota*

Entering the Minnesota state house of representatives marked the beginning of my two-tiered existence. There was the daunting task of having become an inspirational figure who was supposed to serve as a counternarrative to the rising tide of hate. It felt like I spent 90 percent of the year 2017 in rooms where people begged me to answer the questions "What do we do next?" and "Will we be okay?"

"That depends on us," I would say.

I'm not sure how terribly inspiring my answer was,

but my role was to create the laws of Minnesota. That was the job I wanted to focus on and had been elected to do. The push for me to address the Trump administration's policies of bigotry was secondary.

Juggling those responsibilities while maintaining my sanity as a human being was the mission of my small office. Whenever we considered accepting the dozens of requests that never stopped coming in every day, we asked ourselves the fundamental question: what would adding my voice or presence to this debate actually do? An event could be extremely important, but I didn't participate unless it made sense for what I was trying to accomplish in terms of immigration, justice reform, or any other of my issues. And the actual work of legislating always took priority, even if it meant I missed out on meeting Cardi B.

That's what happened with my cameo in the music video for Maroon 5's single, "Girls Like You." When Connor told me I had been asked to appear alongside nearly thirty other famous women in the video, I asked him what I did with every other request: "Should we do it?"

"What are you talking about?" Connor said. "Adam Levine! Um, yeah, you've got to do this."

I accepted and arrangements were made for me to go to Los Angeles to shoot the video. The only problem was that as the date approached, four of our bills were scheduled for a hearing on the day I was supposed to fly to the West Coast. There was no choice; we had to work on the bills that impacted real people's lives. The band wound up accommodating me so that I could be in the video though I missed the first day of shooting, where I also missed most of the celebrities like J-Lo, Mary J. Blige, Camila Cabello, and Ashley Graham (although I did get to meet Ellen DeGeneres!).

**ALTHOUGH A MUCH WIDER WORLD WAS INTERESTED IN PAYING** attention to us, we still remained the Minneapolis organizers who were excited to be in the state house to push the envelope. We had led a movement by challenging a forty-four-year incumbent, and now that the battle against Kahn was over, I had to figure out how to be a colleague to her former colleagues.

It was no easy task. Unseating an elected official with that long a record is messing with legacy. Kahn, the longest-serving female elected official in the United States, had been a powerhouse. Everybody lived in awe and fear of the power and influence she had built over more than four decades. If you misbehaved by crossing her, she was going to make sure you felt her disappointment, either at the ballot box or in failed legislation. So everybody was shocked when I did the unthinkable and took her out. On top of that, I didn't suffer any consequences for doing so but was actually celebrated by Minnesotans.

The result was that some legislators had an existential crisis. Kahn was an institution. If an institution can be dismantled by a girl from nowhere, what did that mean for them? A lot of longtime incumbents questioned what their power was if new candidates didn't fear challenging them and voters didn't miss them when they were gone.

Their anxiety wasn't unfounded or silent. As soon as I did my rounds, colleagues explicitly expressed their fears that I was going to recruit candidates against them. It was no secret that I was irked with how complacent some legislators had become to certain injustices. I wanted those of us who had been elected to act with a sense of urgency. As far as I was concerned, if you weren't constantly worried about the work you were doing on behalf of the people you represented, then we had a problem.

Legislators weren't the only ones who thought they knew what I was going to do when I got into office. The grassroots infrastructure we had created during my campaign had given rise to a team of young people who, energized by their first taste of political organizing, didn't want to let go now. They were ready to dismantle more incompetence by challenging other status quo candidates.

I had to manage expectations from my base while I figured out what success meant in a space that was hostile to my presence. Everyone expected that I would internalize the hostile attitude the kingmakers and operatives were directing at me. But I have never allowed myself to let anyone make me feel that I didn't belong. Perhaps it's because I don't have a fear of rejection. I was an equal member of the caucus, elected by the people, same as them.

So what did I do? What I always do: dive in. On the first day of our caucus meeting, I decided to run for leadership.

**STANDING BEFORE THE REST OF THE DEMOCRATIC MEMBERS, I** said I understood their fears and anxieties but that fundamentally I was there to do the work. I could put my skills to good use in recruiting strong candidates and fundraising with our grassroots coalitions to put our caucus, currently in the deep minority, in a position where it could win back the majority.

I didn't beg or plead; I believed in myself and made my case for why I should become the assistant minority leader. Then my five minutes were up, and they handed out the secret ballots. Everyone but two people voted for me.

After we were sworn in, a staff member of another rep relayed to Connor that his boss was not happy I had been elected to leadership. "They made a mistake," the staffer said. "You elect people into leadership who are loyal to the caucus." This legislator had great anxiety that I was going to be disloyal to *him* by actively recruiting against him.

I could have waited for other members of leadership—no doubt hearing the same complaints relayed to Connor—to explain to me that what I was doing was wrong. I'm not a passive person, however. Waiting for others to come to you feeds into a dynamic where you don't feel equal and only increases the tension.

Instead, I decided that if the other rep didn't have the guts to talk to me directly about his concerns, I would give him the opportunity to do so. I went to go find him in his office. "I hear you are afraid I am going to unseat you," I said. "I don't know much about your district or what you've done in it. But if I was going to find someone to run against you, I would tell you to your face and wouldn't think twice about it."

My fellow legislator responded by listing his accomplishments in his district.

"That's great. Next time you have a problem, you know where to find me. Here is my office number," I said, handing him my card.

As I was walking out, he said, "I have been racking my brain to figure out what's different about you and I got it."

"My hijab?" I said sarcastically.

"No, it's that you somehow walk in like you're a man."

"A white man."

"Yes, a white man."

Constantly being underestimated has always been helpful to me. It was really hard for him to imagine that a ninety-pound refugee hijabi could be confident enough to walk into a leadership role of a caucus whose most important incumbent she defeated. What did he expect? For me to walk around with an inferiority complex? I didn't feel inferior to anyone in the state house. Having been elected by the people, I had just as much right as anyone else to be there. (The rep and I didn't have any problems after that, although he also decided to retire instead of running for reelection when his term was up.)

My brand of optimism is based on my denying myself any sense of victimization and taking comfort in the fact that whatever difficulties present themselves today, they will not exist tomorrow. I believe that by pushing hard enough, you will eventually end up somewhere better. Some have observed that I have an "iron spine." I prefer to see it as a process of figuring out how to channel every challenge into an opportunity. That mentality, which worked in the state house, has always worked for me.

HAVING WON KAHN'S SEAT WAS, ON RARE OCCASIONS, HELPful. There were members who did not like her, especially on the Republican side of the aisle. Some of my Republican colleagues perhaps gave me a little deference because of the way I had defeated an entrenched incumbent. Others viewed me as a potential ally in pissing off the Democratic establishment. Either way, I worked on building those relationships as a way to overcome the challenge of getting Democratic legislation through a Republican-controlled house.

As Connor laughingly put it, I used to "wander off." He was describing how I spent almost every minute of free time in my schedule lobbying other members. Showing up at people's offices, I was always fascinated to watch how people who weren't expecting me reacted to my arrival. When I knocked on their doors to ask if they'd talk to me about my bill, they didn't know what to do with themselves.

I found legislative partners in those who weren't threatened by my passion—like Larry "Bud" Nornes. I sat down with the Republican, who had represented his district in the west-central part of the state since 1997, because he was chair of the Higher Education Committee. I had a bill that addressed funding for parents working toward earning a higher education.

We had a values-based discussion that started with the fact that we both agreed that we want people to be able to get an education while having children. However, we know it takes longer to get a degree, because most parents can't go to school full time. I proposed legislation that provided more money for childcare and extended the time period a student could use it past four years. Part of the bill was to include not just undergraduates but graduate students as well for those who could qualify for the grants. This was a stand-alone bill with my name on it, but I didn't care about getting recognition for it. I cared about it getting a hearing. So I pitched the idea of his pulling my bill and incorporating it into his committee's bigger package of new laws. "This is good legislation," I said. "You can put it in your omnibus, so long as it gets passed."

It wasn't just one conversation and my lobbying wasn't just with Nornes, but at the end of the 2017–2018 Minnesota legislative session, the higher education omnibus bill included my bills for the expansion of childcare grants for student parents, the introduction of a bill of rights for students who are parents or pregnant, and a summer program.

The media made a lot of my unlikely friendship with a lawmaker whose Republican roots went back to his great-grandfather, who was a member of that party who served in the Minnesota house in the early twentieth century. There was no magic to our connection. I paid attention to what he talked about in committee and made a connection to him on a personal level. When reporters asked Nornes why we worked so well together, he said I was a hard worker and knew my stuff.

The question was a fair one. I was someone who came in with a reputation as being the polar opposite of everything the GOP stood for, so it was hard to understand why the Republicans would be willing to work with me. Within my committees on higher education, state government finance, and civil law and data practices, we had a level of achievement unusual for longtime house members—let alone a freshman.

Any success I had in the state house began with real

conversations, which was directly couched from my childhood. I grew up knowing that hard things only get harder when you don't have real conversations about them. Whether it was loud or not, whether you cried or laughed, we always talked it out. That didn't mean over-talking. Even when we were screaming at one another, it wasn't to devalue someone's pain or concerns. Everyone has a right to their thoughts and feelings. Everyone has something to say, even the youngest child.

That belief, drilled into me when I was young, served me well—not just with my colleagues, but also with the diverse constituents in my district. There's a truth I've learned since my days as a community organizer that remains just as true today: ultimately, the people who have the problem have the best solutions to it, if they are given the space, time, and opportunity to work it out.

Whether it was LGBTQIA+ activists fighting against conversion therapy, teens of East African descent starting up a secondhand clothing store as part of a youth entrepreneurship program, or members of the Machinists Union hit with repercussions from President Trump's trade war with China, the fundamental process was always the same as it was with striking legislative deals in the House. Listen up, pay attention, and get to work.

# 19

# RUNNING AGAIN

*2018–2019*

*Minneapolis, Minnesota, to Washington, D.C.*

In the early summer of 2018, I was in the hospital where my father was recovering from surgery when Keith Ellison called me on my cell.

Keith was the U.S. congressman and deputy chairman of the Democratic National Committee. When voters from Minnesota's 5th congressional district chose him to represent them a decade earlier, he was the first Muslim elected to Congress, making him the country's highest elected Muslim official. Keith, a fellow outspoken progressive, had been a mentor to me in many ways. I answered my phone.

He was calling to say he was running for state attorney general, which would leave his congressional seat open. I was one of a handful of candidates he had contacted to discuss running during the November midterms.

Elected to office less than two years earlier, I hadn't yet given thought to the next step in my political career. Maybe I would have considered running for Congress at some point—ten years from now, not in ten weeks!

I didn't have a desire to embark on a new campaign. It had been a long year, and I was exhausted. Before I had even hung up the phone, I knew the idea was not feasible. For starters, I had an absurdly short time to make the decision, since the filing deadline was in thirty hours. To put that in perspective, most congressional candidates take at least three years to put a team together and prepare for a real campaign. I didn't know if I could even assemble a team at this point. All the great political organizers I knew were already either working for other campaigns or taking a break from the ones they had just mounted. Last but not least was the daunting task of raising upward of $1.5 million in less than three months.

On top of all the professional logistics of mounting a last-minute election campaign, there were the personal issues. I had spent the legislative year arriving at the state house around seven A.M. every morning, not returning home much earlier than nine P.M., and promising my children that soon I would work less. The demands of my family—which now included caring for my ailing father—made the whole enterprise even more impossible.

No one expected this, not me, not my staff, not my family. But as soon as my name was floated as a possibility to a local political organizer, Wintana Melekin (who later became my political director), my candidacy became a foregone conclusion. It wasn't long after I hung up with Keith that my phone started to blow up with calls and texts from friends, family, colleagues, political operatives, and the media.

The only message I remember was the one from Connor.

"It's obviously up to you," he texted, "but if there's any part of you that wants to do this, I think you should do it now. And I'm fully on board. Let me know if I can help."

All I texted back in reply was a question mark.

?

The next day, I sat down with my staff to discuss the open seat. I texted a few people I trusted to see if they wanted to join, and two hours later more than thirty-five people showed up. The adrenaline and excitement and planning were instant. People were taking up roles and writing checks before I even had a chance to talk it over with my family.

In fact, my family members were the last to learn about my decision. Isra found out about it from her teacher, who told her in class, "Your mom just announced she's running for office." Like Isra's teacher, Ahmed saw the news on social media. There was no time to stop and make calls. That's how fast everything happened. Not a single person in my family was happy about it, either. (And the only person happy about the decision since then is Aabe, who on my swearing-in day told me a Somali saying that roughly translates to "If I die tonight, there is nothing left for me to see.")

The way the infrastructure for my congressional campaign manifested itself was perhaps what most persuaded me to go through with it. Whether online or on the street, people in the district made their strong desire for me to run clear.

On June 5, 2018, I filed the paperwork to run for the United States House of Representatives, telling the crowd of media that had come out to cover the bureaucratic moment, "I'm excited to go and be a voice for the voiceless in the Capitol."

**THE NEXT TEN WEEKS UP TO THE PRIMARY WERE THE FAMILIAR** storm of door knocking and meet-and-greets. It was during one of the latter events that an attendee said to me excitedly, "You'll be the first person in Congress to wear a hijab!"

The comment was meant as a vote of support, another happy first. But I heard an ominous warning. Later, I repeated the hijab remark to a member of my staff, who confirmed my suspicions about covering my head in the Capitol when she said, "I wonder if that's allowed." I looked it up, and sure enough, headwear of any kind had been banned from the House floor since 1837.

As soon as I saw the ban, I felt dizzy, like I had just been punched in the stomach. "How in the world am I going to make this work?" I wondered, quickly spiraling into thoughts of the worst-case scenario: I win, but I can't serve.

Once I was aware of the possible obstacle, I couldn't get my mind off of it. At the same time, I didn't know what to do. I couldn't call around to Democratic leaders to figure it out, because I didn't know if I was even going to win the primary. And if I did, who was to say that Democrats would win a majority of the House? I couldn't imagine it would be very easy to get an exemption for my head covering if the Republicans were in control. Every day, I watched the crisis inch closer like an accident waiting to happen.

When the hijab became a permanent part of my practice in 2005, it was not to make a statement for others by living visibly as a Muslim but rather to anchor myself internally to a set of values concerning how I wanted to exist in the world. As with all decisions that are lived every day of one's life, it took on added meaning with time.

Public associations with a faith often eclipse a person's individual relationship to that faith, especially if it's one as scrutinized as Islam has been since 9/11. In my community, right after that horrific day when almost three thousand people were killed by Al-Qaeda terrorists, women took off their scarves and men cut their beards. There was an imminent threat to being visibly Muslim as we reckoned with the automatic association with Islamic terrorism.

At that time, it was important for me to wear my hijab so others could associate a normal young woman, who worked and went to school, with being Muslim. I wanted to subtly shift the narrative simply by acting as a positive ambassador. As I quietly studied from a textbook from under my hijab in Perkins, I was saying to everyone else in the coffee shop, "Look! We aren't chanting 'Death to America!' or blowing up buildings. We are your neighbors."

Living authentically is the best form of resistance. When I got to the state house and experienced discrimination or was subjected to derogatory remarks, I didn't call out elected officials or report it. I didn't do anything other than act as a visibly Muslim equal. Sometimes I was a little more visible than others, as when one legislator, assessing my outfit, pointed to my skirt and boots and said, "I get this." Then he pointed to my hijab and said, "But I don't get this." My response was to wear a full black abaya, complemented by a black hijab, while we were in session that week. My outfit, which caused quite a stir in the bitter cold of a Minnesota winter, left the legislator dumbfounded. I could have told him off in the moment he commented on my appearance, shrunk, or did what I wound up doing—showing him that I wasn't asking for his approval. (I wanted to wear the abaya with a niqab to Trump's State of the Union, but my staff worried I would be arrested by the Secret Service.)

And yet I also rail against having every action I take reduced to a social construct stemming from my religion, stripping me of the complexities of multidimensional thought. I am a human, not a figurehead. I have always chafed at owning other people's notions about my identities, be it what it means to be a mother or a member of Congress.

**ONE OF THE MOST TOXIC MISPERCEPTIONS OF MY FAITH IS THAT** because I'm a Muslim, I hate Israel and the Jewish people. Although that couldn't be further from reality, whenever I criticize Israel, it is filtered through this lens.

Case in point was my response, on February 10, 2019, to a journalist's tweet about House GOP leader Kevin McCarthy, who said two days earlier that he wanted to punish me for my views on Israel policy. "It's all about the Benjamins baby," I tweeted in a reply, using a Puff Daddy lyric to comment on the power of moneyed influence in our country—in this particular case, the American Israel Public Affairs Committee (AIPAC) on our foreign policy in Israel.

As it turned out, it was not clever but hurtful. I immediately began hearing from friends and colleagues who explained why the tweet was offensive. As soon as I understood the history, I quickly apologized for using what I learned was an age-old anti-Semitic trope about Jewish control through money. I want the ability to be heard and have my full humanity recognized. In return, I do everything I can to make sure that is true for others, too. So I apologize when I denigrate others or make anyone feel invisible.

"Words carry historical baggage," Peter Beinart wrote in the *Forward*. "That doesn't mean it's illegitimate to talk about AIPAC's fundraising, any more than it's illegitimate to talk about O. J. Simpson killing a white woman. Given the toxic stereotypes that such discussions evoke, however, they must be handled with care. Ilhan Omar didn't do that. Which is why she was right to apologize."

I didn't just apologize, a month after the initial tweet, I wrote an op-ed in *The Washington Post* where I tried to clarify my position on the Israeli-Palestinian situation: "My goal in speaking out at all times has been to encourage both sides to move toward a peaceful two-state solution."

I asserted that the founding of Israel was "built on the Jewish people's connection to their historical homeland, as well as the urgency of establishing a nation in the wake of the horror of the Holocaust and the centuries of anti-Semitic oppression leading up to it." But I added, "We must acknowledge that this is also the historical homeland of Palestinians. And without a state, the Palestinian people live in a state of permanent refugeehood and displacement." In conclusion, I wrote, "A balanced, inclusive approach to the conflict recognizes the shared desire for security and freedom of both peoples."

What people made out of my original comments, though, couldn't be undone. More than a year later, Bernie Sanders—a Jewish candidate for president—was attacked by the right wing for running an anti-Semitic presidential campaign because he had accepted my endorsement. I'm asked all the time in all different places to relitigate the issue. At a meeting of local Minneapolis LGBTQIA+ activists to discuss the community's agenda for 2020, one participant brought up the question of my support for Israel. I wasn't tired of trying to make my case. "I'm not here for your support," I said. "I'm here to support you, so let's talk about what you need."

You can't take away the past; you can only add to the narrative. There is a narrative about Muslims that already exists. I'm not here to undo or rewrite history. That is propaganda or an impossibility. What I, and others, can do is expand on the notion of what it means to be Muslim, continue the story line that survives alongside us.

**WE ALL HAVE OUR BLIND SPOTS. AFTER THE INCIDENT IN THE** Washington, D.C., cab where the driver suggested I had a bomb under my hijab, someone on social media tried to apologize to me on behalf of "white Americans." But I had to check that person's assumption, because my driver that day wasn't white. In fact, he had an African accent that made him sound like some of my relatives.

This world can be a confusing place. I have been accused of being an anti-Semite by American media and politicians and an agent of the Muslim Brotherhood in an Islamist plot to overthrow Congress by Saudi Arabian network news. That is why I work hard to defend my ideas, but my identity is not up for debate. My ideas and policies are the product of a rational and thoughtful deliberative process. My identities are just one plot point in a long and crazy story.

Freedom of thought and speech—the essence of what it means to be human—is my right no matter the color of my skin or my religion. It's the right of people who speak with accents or whose hair texture is different. Religious minorities, the formerly incarcerated, those without bank accounts or homes, the neurodiverse. No label should rule out participation.

All that is true, and yet I'm learning that there are

times where, despite my best intentions, my external aspect serves more to hurt than help a conflict. From the time I could walk until the present day, my instinct has been to jump into any battle where I feel as if I can do something (and I always feel as if I can do something). If I'm honest with myself, there are situations where my trying to help causes more harm than good.

I am, by nature, a starter of fires. My work has been to figure out where I'm going to burn down everything around me by adding the fuel of my religion, skin color, gender, or even my tone. Knowing not just yourself—your personal motivations, beliefs, and psychology—but also how the world interacts with you is vital to true and lasting progress. That doesn't mean you shrink away. Sometimes it's enough to quietly show up to lend your support. The politics of "moral clarity and courage," which I often reference, includes lending one's voice *and* listening.

There is no way to do the kind of work I do, to have the honest dialogues that lead to solutions to constituents' issues, without bumping into things and hurting others. That's just human nature. Ideally, though, I remedy it. While not popular in the Trumpian vision for America, introspection and contrition are signs not of weakness but of strength.

My hijab is a personal reminder of the tension between submission and struggle. But after winning my congressional primary in August 2018, practically assuring that my firmly Democratic district would send me to

Washington in November, I wondered if it would keep me from serving in the House.

**MY MIDTERM PRIMARY ELECTION WAS AN UNMITIGATED SUC**cess, although I had had my doubts along the way. The experience of running for Congress couldn't have been more different from my first election, starting with the formation of my campaign. My run for the Minnesota house of representatives unfolded slowly and in a way that was somewhat similar to the work I had done organizing around particular issues, like the $15 minimum wage. I had consultations on the pros and cons of my running, collected data, talked to different stakeholders about my methodology. I had predictors in place for my success—and my failure—in order to guard against them.

For the congressional primary, we had a little over thirty hours to decide to run, come up with a fundraising plan, and put together a campaign team. It wasn't just the time pressure that was different but also the political landscape. I was running in Trump's America. Part of the reason I had been asked to run was my national profile, which arose out of my battles with our president. That notoriety gave me name recognition for good and for bad. We had to deal with smears, overwhelming media attention, and the challenges of keeping the focus on the issues that had convinced voters to cast their ballots for me the first time around.

To this day, I am shocked by the way my team—a group of people who had never been part of or experienced a congressional race—went from zero to serious contenders in no time at all. Joelle, who ran the campaign; David, my campaign chair; Connor, my scheduler; Akhi Menawat, my communications director: all firsts and all with a combined average age of no more than thirty.

Some things *didn't* change, however. I ran on national versions of the platforms I had promoted in Minnesota's legislature, such as automatically registering citizens to vote when they turn eighteen and letting more rather than fewer refugees into the United States. I also put in sixteen-hour days in the middle of June calling hundreds of delegates to try to persuade them to spend Mother's Day at a caucus instead of brunch—and, of course, to endorse me. I got pneumonia (I blamed the air conditioning in our office; I'm allergic to air conditioning and never use it if I don't have to), but we won our Democratic endorsement at the convention within ten days of announcing.

In the primary, I won my district by more than 20,000 votes over the closest candidate to me in the polls. The race also broke a record for voter turnout in a midterm primary, with 135,318 ballots cast for Democrats. *The New Yorker* compared my stats to Alexandria Ocasio-Cortez's historic upset in New York's 14th congressional district, which she won by 4,000 votes out of only 28,000 cast. In both districts of roughly a little more than 700,000 people, it was a good day for democracy.

While calls of congratulations from D.C. started to come in, I set figuring out how to actually sit as a member of Congress on the top of my to-do list. In my phone call with Congresswoman Pelosi, gearing up to retake the House during the midterms, she said, "That's nothing. We'll get it done."

I wanted to say, "But what if we don't win the majority? What if you're not the speaker?"

Some call it a refugee mentality—the psychology that remains long after you've been forced from your home—but I am always bracing myself for the bad that comes with any good. My survival is dependent on not being blindsided in a world where nothing is constant. Basically, I demand constant vigilance from myself. No, it's not a relaxed existence.

Hijab-gate consumed me. While I lived in fear of the press getting ahold of the issue and firing up the right wing to mount a campaign, I talked about it incessantly with those in my immediate circle. Was it covered under the constitutional right to express my religious freedom? If the GOP blocked me from wearing a head covering, would I have to sue Congress? I didn't have money for that kind of legal fight. My mind spun with all the permutations of what could go wrong.

I argued with Aabe, who sided with Pelosi that it would "get done." Whether I'd be able to wear my hijab on the House floor was just the latest iteration of a long debate between us about the American system. Aabe believes things ultimately work out for everyone who wants them to. I thought that was naive from the minute I arrived in this country and saw homeless people on the ride into Manhattan from the airport. Having witnessed and played a part in the inner workings of the system as an elected official only strengthened my position. I ran for government precisely to challenge the systemic injustices faced by those perceived in our society as not worthy.

**IN SEPTEMBER 2018 I TRAVELED TO D.C. TO MEET WITH A HOST** of people, including Representative Jim McGovern of Massachusetts. McGovern was slated to become the chairman of the Rules Committee if Democrats won the majority, and would be responsible for drafting the provision to carve out a religious exception to the ban on hats on the House floor.

In the blue wave of the 2018 midterm elections, during which Representative Rashida Tlaib of Michigan and I became the first Muslim women elected to Congress, the Democrats did win the majority of the House seats. Still, I worried. First we had to figure out what the new rules package was going to look like. The provision that would allow me to wear a hijab in the chamber was part of a much larger group of provisions, some of which I wholeheartedly disagreed with.

For example, there was a proposal to require a three-fifths supermajority to pass any legislation that raised income taxes on the lowest-earning 80 percent of taxpayers. That was an improvement on the Newt Gingrich–era requirement of a supermajority as opposed to a simple majority for *any* bill that included a tax increase. However, I and other members of the Congressional Progressive Caucus opposed it on the grounds that it would make initiatives such as Medicare for All nearly impossible to achieve. Would my conscience allow me to vote for a package that allowed me to wear my hijab if it meant the death knell of important progressive ideals?

Thanks to Jim McGovern, the hero of the story, I didn't have to choose between self-interest and the greater good. I spent time with him and his staff, who did the big lift in figuring out the language and compromises necessary to find a package we could all live with (including getting rid of the supermajority amendment).

So it was that on January 3, 2019, I wore my hijab as I was sworn in en masse with the rest of the members of the 116th Congress, which included a record 101 women legislators, 43 of whom were women of color.

Speaker Pelosi and McGovern not only made sure I was able to be on the floor to vote for the package, they also had me author the amendment that overturned the 181-year-old ban on headwear to allow for hijabs, yarmulkes, or head coverings for illness and hair loss.

I got emotional at the suggestion because of what putting my name on the provision symbolized. They weren't making an accommodation for me, but were encouraging me to take ownership of the change I had sought. It said a lot about their character.

Naturally, not everyone found the historically diverse House a cause for celebration or was happy about the new rule regarding headwear. One Christian pastor complained the floor of the House "is now going to look like an Islamic republic."

"Well sir," I replied on Twitter, "the floor of Congress is going to look like America . . . And you're gonna have to just deal."

# 20

# THE WORLD BELONGS TO THOSE WHO SHOW UP

*2019–2020*
*Washington, D.C.*

Before I was sworn into Congress, I spent a day on Capitol Hill settling into my D.C. office. A room had been set up with small consultation tables for new members going through the same process. Each table had a different function. One was designated for setting up the health insurance plan for you and your staff. At another, you could pick out furniture. There were tables to sign up for classes and get pictures taken for ID badges, and even a table to figure out the process of the other tables.

The last table Connor, now my chief of staff, and I arrived at, was the Capitol Police, who were responsible for members' security in their districts, homes, and congressional offices. We sat down across from a white man, who began to instruct us on protocols and best practices in regard to issues of safety. Except he wasn't really instructing *us*; he was clearly talking to Connor. Although I was directly across from him, he refused to make eye contact with me. In fact, as I leaned in to try to hear better, he turned his chair so that he was facing only Connor.

Connor is not a fan of confrontation. That's usually my job. Plus, he was deferring to me, waiting to see how I wanted to handle this obvious case of mistaken identity. Really, even if the person across the table hadn't gotten a briefing with a picture of me, you'd think he would have realized that someone named Ms. Ilhan Omar wasn't a white man.

I decided not to say anything.

Meanwhile, Connor shifted his chair to open up the conversation to include me, his boss. But that only made the man lean closer into Connor and give me more of his shoulder. "What do you think?" Connor asked me, trying a different tack. Again, the man leaned in closer to him *and* lowered his voice.

By this point, I was trying hard not to laugh. It wasn't just funny to see Connor so uncomfortable. It was also ironic that someone charged with keeping me safe didn't even know what I looked like.

"Good luck!" The man shook Connor's hand heartily, and just like that we ended the ordeal and left the table.

I understand why the man had assumed Connor was the elected official. There has never been a member of Congress who looks or sounds anything like me. It was a fascinating contradiction to both stand out and completely fade into the background at the same time. But being a first is all about contradictions. I had been catapulted to international fame in part because I was the first Somali American legislator. Firsts are more than just iconic; they are totems for the groups they represent. The Somali community saw their fate intertwined with mine, which is why they were often my harshest critics when I stumbled. In turn, I inherited their collective failures. Any time a Somali person was arrested, I was tagged on social media. That aspect of being a first is quite painful.

Whenever I am asked which famous person dead or alive I would want to meet if I could, my answer is always without fail Margaret Thatcher. It surprises people that the leader of Britain's Conservative Party is my greatest shero. While her politics aren't mine, she was also a first—the first female prime minister of Britain. Thatcher was a self-starter in the grandest of ways.

Without any kind of special invitation or connections, time and time again she showed up in rooms filled with men and didn't have to do much to lead them to decide that she should be in charge. That she held office for eleven years, longer than any other British politician in the twentieth century, proves her level of confidence and intellect. It was her fearlessness and internal sense of equality, however, which made her able to pull off being a first.

A month before summer break after ninth grade, I got into the habit of wandering into the math class for English-language learners toward the end of the period. I would hang out in the back of the room, which was where the Somali boys sat. We goofed around and chatted about Somali politics. I don't know why the teacher let me do it, but he was a friend of my father's. In fact, any time Aabe and the teacher ran into each other in the coffee shop, Aabe would ask if the teacher had seen me. Dad, concerned about my life as an American teen, was looking for intel. The teacher told him not to worry. "She's an Iron Lady," he said. "If it were Ilhan against a hundred men, I would worry for the men."

Aabe still calls me Iron Lady, because he thinks I have no fear. I was able to convince people that I can lead without any credentials or validation. My strength doesn't come from a lack of fear but from an overpowering sense of moral outrage. I stood up to the bully in my elementary school class only because when he was picking on the weak child I lost sight of the large boy's size.

"This really was a victory for that eight-year-old in that refugee camp," I said on election day 2016. "This was a victory for the young woman being forced into child marriage. This was a victory for every person that's been told they have limits on their dreams."

I want to help all those who feel small to feel large; to give strength to all those who believe they are weak; to make loud those who think they are voiceless. To me, that is the American dream.

**WHEN WE FIRST ARRIVED IN THE UNITED STATES AND AABE RE-**plied to my bafflement at the trash-filled city we had landed in by saying, "This isn't our America. We'll get to our America," he meant, "This isn't where we're staying. We'll get to the city we're going to."

But at the time, to me it meant we had arrived in the wrong *America*. We were eventually going to get to the right America, though, the one that matched the image in my head. That is still what it means to me today, and I'm still on the journey to find our America.

Although it might not be the reality every day for everyone in this country, the American dream isn't just something immigrants talk about who are coming to or want to come to this land. It is part of the American psyche and ultimately what we citizens of the United States are all searching for.

We all want that abundant turkey dinner in that beautiful home. We all want those welcoming schools and neighborhoods where kids ride their bikes or play together. We want to export that image to the rest of the world, because that is ultimately what we want for ourselves.

I don't know if those videos I watched in Kenya before I boarded the plane for the States are still run in orientations for new Americans today. I do know that the majority of refugees and immigrants who arrive here from all over the world will most likely never get to have as beautiful a home, as nice a school, and as safe an environment for their children as those that were portrayed.

That is both a terrible and beautiful fact. To be part of this constant aspiration for a better society is gifted to you as you get on the plane. Drive and hustle are as important to the American dream as cornfields and white picket fences. It fuels new Americans constantly working their butts off in search of the promise.

Although we're nowhere near it, the work toward a more perfect union was enshrined in our Constitution at the founding of this country. I worry, though, that the striving aspect of the American spirit is greatest in our new immigrants, because many of us already here have become complacent.

We are not living up to the ideals we export to the rest of the world. In our country, we've normalized inequities and hardships to the point that we don't even recognize them as such. In the past, we didn't talk much about racial, gender, cultural, religious, and sexual biases, because we're all supposed to be equal—the only differences among us being our ability to work hard and persevere. But not talking about the underlying economic structures that allow some people to amass ever-increasing wealth while others struggle to feed and house themselves serves only those at the top.

What is missing in the conversation about our national identity is that there exists an elitist system belying the vision of ourselves as rugged individualists free to pioneer whatever frontier we can find, from computer codes to movie scripts.

In our cultural mythology, there is a false sense of the power in and of personal achievement. Hatred has built up in our country in the form of identity politics, but I believe that most of it masks economic anxiety. When people suffer from college degrees that don't bring jobs but only huge student loans, or unsustainably high premiums for insurance plans that still leave them bankrupt after a catastrophic illness, they understandably might hate others with fewer money problems. The anger is misplaced on individuals. It should be directed at the way society is set up. Then it should be used to change the way society is set up.

I truly believe the America that we all want and deserve is not just a myth but a place that could actually exist. But only if we work for it, and not just individually—not just for ourselves, our families, or our ethnicities, religions, or other groups to which we belong. We need to work for it together.

The essence of community organizing is finding common goals among varied individuals and using the collective numbers to advocate from a place of strength. It's the next step in the process of representative government. Once you've convinced yourself about the importance of voting, then you figure out: What do I care about? Who cares about these things, too? And how do we move them along?

As we continue to perfect our union, citizens, neighbors, coworkers, and family must keep expanding our circles of self-interest to learn and relearn the fundamental truth that we are all connected. The more invested we are in one another, the better all of us ultimately will be. This is the philosophy of interconnectedness that I operate on as a legislator in a country where there is enough abundance to achieve all our goals.

It is the opposite of the myth of scarcity, where what's mine necessarily takes away from yours. We become obsessed with who has more and depressed about all that we lack. This mentality is what pits minority groups against one another in a fight for scraps. Those propping up the status quo are happy to see us so distracted. I would like to reframe the old adage that one person's gain is another's loss. I want your loss to be my loss; your gain, mine, too.

While this might sound utopian, it's actually a practical argument. Consider wealthy communities and their biggest fear, which is typically around safety. Who is a threat to their safety? It's the person unable to make a decent living. If there are mass foreclosures, those without homes are going to need services that we all wind up paying for. The same is true for healthcare and education. It's an old story.

Shifting the framework of how we approach challenges in our society from *I* and *they* to *us* begins with individual points of compassion. The more we listen to and learn from those with different backgrounds and present circumstances from our own, the more we can find connections to our own lived experience.

Although the win-loss structure seems impossible to break in our political system, nowhere is empathy more important than for those of us in office. I've seen it up close as a new American, a community organizer, and a member of Congress: we can't eradicate problems unless we put ourselves in the shoes of those impacted by the solutions we implement.

My aha moment in realizing that government representatives must be fluent in the day-to-day struggles of those they serve came when I was working as a nutrition educator for underserved communities in Minneapolis. Here I was, telling struggling immigrants the properties of vitamins found in fresh strawberries; meanwhile, they had no way to get anything close to a fresh strawberry. Or how about the young high school moms whom I had been instructed to counsel on calcium intake. They couldn't concentrate on my powder demonstration because they were too busy worrying about catching the bus with their babies. Those in the business of creating the policy that I was implementing had no connection to food, transportation, or childcare insecurity.

We don't have to be hungry or gay or black or anything else to understand the needs and wants of those who are. We just need to have conversations with one another and know that every policy is personal because at some point, if you widen the circle out far enough, it will touch you.

The democratic process is like a baby learning to walk: there's a lot that goes wrong, and a lot that goes right. Mostly, though, the process takes a long time. I worry that people have lost sight of the flexible thinking and tireless efforts that have gone into the progress in science and art, and the physical and social infrastructure we've made over the history of this country.

Recognizing my psychology as a refugee who has seen her home devolve into chaos basically overnight, I still feel it's my duty to call out the lack of awareness about the disintegration of civilization that is possible anywhere. One day you are going to school, to work, shopping in the market, eating in restaurants with friends, and suddenly, the next day you wake up to complete darkness. In reality, that doesn't happen overnight. There is always a process of corruption and abuse that erodes the systems we depend upon. But it can happen only when nobody is paying attention or people stop caring. That's how we regress.

**THE PRIVILEGE THAT IS EVERY AMERICAN'S BASIC RIGHT TO** participate in the democratic process was imprinted on me from the moment I accompanied Baba to his first caucus. In the crowded room in Minneapolis's 5th district, everyone was buzzing with excitement. But I quickly saw how some, like the new Somali immigrants, couldn't fully access the process because of language barriers or lack of information about the strange events unfolding around them. The unfairness ignited me to action, as it did Baba.

After he had participated, I couldn't find my grandfather—until I looked back in line and located him instructing those who only spoke Somali on what the ballot looked like when they got to the front.

On our way home, he talked to me at length about how the whole thing could run more smoothly next time. "We'll read the material on each candidate and proposal before we go," he said, "and we should get those signs about 'Vote Here' translated into Somali . . ."

Over the years, every time I took him to vote, it was the same thing. He left with an agenda of tweaks to improve people's understanding of voting and access to it. His striving for a more perfect political process was a constant. It annoyed some people, but I appreciated how he inserted himself in places others thought he had no business offering an opinion. That was just his nature to say, "Let's figure out how to make this happen and how to make it happen *better.*"

I'm proud to claim Baba's bossy nature as my inheritance. Ensuring that the democratic system is prepared for all the challenges it faces ignites me to this day. From increasing access for new Americans united in their excitement to participate to reaching out to the disillusioned to explain the purpose of their involvement, those who are conscious of what isn't working have a responsibility to figure out how to make it better. But it all starts with just showing up, and that's something all of us can do.

# ACKNOWLEDGMENTS

*Mahadsanid* to everyone who helped turn a refugee's story into a book. A big thanks to my agent, Steve Ross, and the talented team over at Dey Street, including Rosy Tahan, Shelby Peak, Mumtaz Mustafa, Anwesha Basu, Kell Wilson, and Carrie Thornton. To my editor, Alessandra Bastagli, thank you for believing in this project from our first meeting. And to Rebecca Paley, who partnered with me in the writing process with patience, grace, and humor.

To my tireless congressional staff in Washington, D.C., and Minneapolis—my story wouldn't be anything without the heroic work you do on behalf of our constituents.

To the organizers, activists, and table shakers who inspired me and helped me build two historic campaigns.

To my congressional guardian angels, friends, and colleagues, who help me courageously fight for a better America.

To our ancestral firsts, thank you for paving the way and cracking many ceilings.

To my sisters, brothers, aunties, uncles, nieces, nephews, and cousins, thank you for being a reliable source of unconditional love, and, Aabe: thank you for championing me even when it came at a great cost.

To my best friend, thank you for helping me chase waterfalls and enjoy life as a spring breeze.

To Isra, Adnan, and Ilwad, you are my everything.

To the great people of Minnesota and District 60B—if I am a first, it's because you had the vision to make me one.

And lastly to those fighting for democracy and the right to vote each and every day: never give up.

# NOTES

**CHAPTER 2: WAR**

17 many from disease and starvation: Annabel Lee Hogg, "Timeline: Somalia, 1991–2008," *The Atlantic,* December 2008; https://www.theatlantic.com/magazine/archive/2008/12/timeline-somalia-1991–2008/307190/.

**CHAPTER 4: REFUGEE**

47 the population had grown to 334,000 refugees: Tammerlin Drummond, "Kenya Welcome Mat for Somalis Fraying," *Los Angeles Times,* October 9, 1993; https://www.latimes.com/archives/la-xpm-1993–10–09-mn-43941-story.html.

#### CHAPTER 5: AMERICAN DREAM

53 offered to only a small number of refugees: Laura Hammond, "Somali Refugee Displacements in the Near Region: Analysis and Recommendations. Paper for the UNHCR Global Initiative on Somali Refugees," UNHCR (2013); https://www.unhcr.org/55152c699.pdf.

53 according to UN estimates: Barbara J. Ronningen, "Estimates of Immigrant Populations in Minnesota," Minnesota State Demographic Center, 1999; https://mn.gov/admin/assets/estimates-of-immigrant-populations-msdc-may1999_tcm36–76617.pdf.

61 The white plastic bag from the International Organization for Migration (IOM): "What Is an I.O.M. Bag?" World Relief Durham; https://worldreliefdurham.org/iom-bags.

#### CHAPTER 7: MINNESOTA NICE

81 11,164 Somalis: "Survey: Nearly 1 in 3 US Somalis live in Minnesota," Associated Press, December 14, 2010; https://www.sandiegouniontribune.com/sdut-survey-nearly-1-in-3-us-somalis-live-in-minnesota-2010dec14-story.html.

#### CHAPTER 12: RETURN

149 Al-Shabab had retreated from the city: "Somalia's al-Shabab Rebels Leave Mogadishu," BBC News,

August 6, 2011; https://www.bbc.com/news/world-africa-14430283.

**CHAPTER 13: POLITICS**

160 "His second-place finish was marked by a dramatic turnout from the city's Somali community": Eric Roper, "Kari Dziedzic Wins DFL Nod for Senate Race," *Star Tribune,* December 7, 2011; http://www.startribune.com/kari-dziedzic-wins-dfl-nod-for-senate-race/135143438/.

**CHAPTER 14: CITY HALL**

183 As I later wrote in an op-ed for the *Star Tribune*: Ilhan Omar, "My Side of the Cedar-Riverside Caucus Dispute," *Star Tribune,* February 13, 2014; http://www.startribune.com/my-side-of-the-cedar-riverside-caucus-dispute/245283731/.

**CHAPTER 17: AMERICA'S HOPE AND THE PRESIDENT'S NIGHTMARE**

216 "but we are proud to have come through it together": Ilhan Omar, "Statement from Ilhan Omar," 2016; https://kstp.com/kstpImages/repository/cs/files/Ilhan%20Omar%20Statement.pdf.

217 "You've suffered enough in Minnesota": "Trump: No Refugees Without Local Approval," CBSN Minnesota, November 6, 2016; https://minnesota.cbslocal.com/2016/11/06/trump-refugees-minnesota-local-approval/.

219 "until our country's representatives can figure out what the hell is going on": Jenna Johnson, "Trump Calls for 'Total and Complete Shutdown' of Muslims Entering the United States," *Washington Post,* December 7, 2015; https://www.washingtonpost.com/news/post-politics/wp/2015/12/07/donald-trump-calls-for-total-and-complete-shutdown-of-muslims-entering-the-united-states/.

222 a 67 percent increase in crimes against Muslims: German Lopez, "A New FBI Report Says Hate Crimes—Especially Against Muslims—Went Up in 2016," *Vox,* November 13, 2017; https://www.vox.com/identities/2017/11/13/16643448/fbi-hate-crimes-2016.

**CHAPTER 19: RUNNING AGAIN**

246 "Which is why she was right to apologize": Peter Beinart, "The Sick Double Standard in the Ilhan Omar Controversy," *Forward,* February 12, 2019; https://forward.com/opinion/national/419206/the-sick-double-standard-in-the-ilhan-omar-controversy/.

247 "the shared desire for security and freedom of both peoples": Ilhan Omar, "Ilhan Omar: We Must Apply Our Universal Values to All Nations. Only Then Will We Achieve Peace," *Washington Post,*

March 17, 2019; https://www.washingtonpost.com
/opinions/ilhan-omar-we-must-apply-our-universal
-values-to-all-nations-only-then-will-we-achieve
-peace/2019/03/17/0e2d66fc-4757-11e9-aaf8
-4512a6fe3439_story.html.

# ABOUT THE AUTHOR

Ilhan Omar currently serves as the U.S. representative for Minnesota's 5th congressional district. In November 2018, she became the first Somali American elected to Congress and one of the first two Muslim women elected to Congress, as well as the first woman of color to serve as U.S. representative from Minnesota. Representative Omar and her family fled Somalia's civil war when she was eight. She spent four years in a refugee camp in Kenya before emigrating to the United States. Omar currently lives in Minneapolis with her family.

Rebecca Paley is a #1 *New York Times* bestselling collaborator and coauthor of many books.